International Perspectives on Undergraduate Research

Nancy H. Hensel • Patrick Blessinger
Editors

International Perspectives on Undergraduate Research

Policy and Practice

palgrave
macmillan

Editors
Nancy H. Hensel
Laguna Woods, CA, USA

Patrick Blessinger
International Higher Education
Teaching and Learning Association
New York, NY, USA

ISBN 978-3-030-53558-2 ISBN 978-3-030-53559-9 (eBook)
https://doi.org/10.1007/978-3-030-53559-9

This Palgrave Macmillan imprint is published by the registered company Springer Nature Switzerland AG.
The registered company address is: Gewerbestrasse 11, 6330 Cham, Switzerland

Foreword to International Undergraduate Research

From the United States to the United Arab Emirates, undergraduate research (UGR) is capturing and maintaining the focused attention of university faculty, administrators and especially of students. *International Undergraduate Research* demonstrates why. The thinking required by, and developed through, UGR is valuable not only to those within the university, but increasingly to employers and politicians who want graduates with the capacity to solve issues of current concern to society, entrepreneurs who craft whole new ways of working in the world.

There are common challenges across the book's chapters of inadequate university budgets and lack of country-wide mandate, but the diversity of approaches to deal with these and other issues is a big illuminating feature of the book. For example, numerous chapters view UGR in terms of its capacity to build disciplinary knowledge and Makhanya emphasises the need for UGR knowledge development to align to national socio-economic challenges and priorities. Elshimi, however, provides a contrast of emphasis for Egyptian UGR, shifting from the more common focus on building the discipline to solving the problems of pressing community and environmental issues. In all chapters, however, UGR is seen to result in substantial, sometimes life-changing, skills and attitudes of research that students develop and apply broadly and deeply on graduation.

Blessinger and Hensel note that UGR has been declared a high-impact practice, yet the nature of the impact depends on how effectively UGR is implemented. Sengupta and Blessinger consider UGR in India, Malaysia and Iraq and note that determining the impact of UGR remains a challenge in most countries. However, knowing the nature of, and limitations to, UGR impact is vital to work on context-sensitive ways of improving it, whether through tweaking, adaptation or implementation of new models. There is the possibility of negative impact through bad experiences, such as under-equipped mentors, poor scaffolding of skills needed for research projects in the curriculum or even culturally unaware implementations.

Because of the risk of sub-optimal UGR, various chapters deal with mechanisms for improving the teaching/supervision aspects of UGR. Makhanya provides understanding about professional development and Donnelly, McAvinia and McDonnell focus on peer learning. The formation of UGR societies, following on from the Council on Undergraduate research in the USA, has emerged in regions such as Australasia (Brew and Mantai), but such formation is noted as lacking in the majority of chapters. Readership groups, such as university clusters in one country, will be interested to read factors that have led to the development of UGR societies and factors that have been inhibitive of them. However, the role of such societies on the upskilling of faculty and the long-term impact on student learning is in need of research attention.

While some countries have focused on the model of mentored UGR that relies on faculty research agendas, others are more invested with in-curricular models, and still more countries entertain blended or multiple models. However, there are challenges with evaluating the impact of different models of UGR, and Wuetherick considers the state of play in Canada of mentored and curriculum-embedded models. Mentored models have the challenge of accounting for UGR outcomes not just compared to other models of learning, but with reference to increased resourcing and its direct or indirect costs. A major issue, then, for global UGR, is the sustainability of mentored models when compared to in-curricula models. Hybrid models that, for example, use students with experience of UGR to mentor less experienced students also need to be researched.

Issues of equity, including of who gains access to UGR and who is able to persist, and benefit are vital for determining impact. This is particularly important in mentored UGR, which has limits to involvement and, typically, selection criteria for inclusion. Chng Huang Hoon and Siew Mei focus on the more inclusive model of UGR embedded in the curriculum in Singapore, which provides opportunities for all. However, this model too has challenges for evaluation. The curriculum space occupied by UGR has its own costs, for example in terms of reduced content coverage, and evaluation of impact must take into consideration controversies of content-covering versus skills-based curriculum.

There is a variety of experiences in the book in terms of how long since UGR was first promoted explicitly in the country of focus, with the USA (Ambos) witnessing an evolving clear mandate over more than four decades that has included engagement with, and funding from, the nation's Congress. Ironically, Germany (Deicke and Mieg), the homeland of the Humboldtian model of research universities, has not had undergraduate enrolment until recent times, so the chapter provides insights for a system just starting to promote UGR, as does the chapter on Japanese UGR (Imafuku). New Zealand (Spronken-Smith) has national legislation that calls for close links between teaching and research, yet its university system nevertheless provides patchy opportunities for UGR.

Taken together, the book does not imply one superior model for UGR, but rather the need for creative and thoughtful implementations that are context-savvy and that embrace a willingness to improve and improvise.

As this book is readied for publication in the dark days of the COVID-19 crisis, I commend the ambition and dedication that the authors convey for the development of graduates who will light the way through their contribution to solving each country's and our global problems using the skills and attitudes that they learn in UGR.

Adelaide, Australia
14 April 2020

John Willison

Preface

Undergraduate research is becoming part of the student research experience for students in many countries. Undergraduate research is not a new pedagogy. Its history is often/widely attributed to the work of Wilhelm von Humboldt, who founded the University of Berlin in 1810. Humboldt stressed an educational approach that unified teaching, learning, and research, and the continuing process of inquiry. American universities, as with most universities around the world today, grew out of the German model. For instance, in the United States, the Massachusetts Institute of Technology (MIT) began formally involving undergraduate students in research in 1969. After Professor Margaret MacVicar founded the Undergraduate Research Opportunity Program at MIT, many other US institutions began similar programs over the next few years.

The Council on Undergraduate Research was founded in the United States in 1978 to assist colleges and universities in developing undergraduate research programs and advocating for federal funding to establish undergraduate research programs. While there are common roots for the concept of undergraduate research, there is no consensus within the international community of scholars on its parameters. It is the differences in emphasis, program implementation, funding, and policies that make the study of undergraduate research across the globe both an exciting topic and a resource for future program development and implementation. While approaches to student research vary, there is however common agreement that undergraduate research positively impacts student learning and contributes to career preparation.

The purpose of this book is to explore the implementation of research-based teaching and learning in countries around the world. International collaboration in high-impact, experiential learning is a significant interest of both editors. Patrick Blessinger founded HETL (International Higher Education Teaching and Learning Association) to advance the scholarship and practice of teaching and learning around the world. He has brought together international scholars to engage in cutting-edge research to develop innovative practices in global higher education, including high-impact areas such as inquiry-based learning, education for sustainable development and social responsibility, university partnerships, equity, inclusion, and leadership. Under his leadership, HETL has grown into one of the largest and most active research networks of educators around the world. While Nancy H. Hensel was the executive director of the Council on Undergraduate Research (CUR), she invited practitioners and scholars of undergraduate research from many countries to contribute to the CUR Quarterly and participate in CUR activities. During her leadership, CUR began several international collaborations.

Blessinger and Hensel identified leading educators from 15 countries to contribute to this book. As they identified potential authors, they wanted a range of perspectives on undergraduate research. They also wanted to include countries from every continent (except Antarctica). They approached scholars whom they knew to be involved in undergraduate research and were leaders in their field. They asked them to consider how student research is defined in their country, organizations that might support student research, national policies and initiatives for student research, and curricular models for implementing undergraduate research.

Chapter Summaries

In Chap. 1 Patrick Blessinger and Nancy H. Hensel discuss how the demand for global higher education of all types has increased significantly over the past few decades and how this has prompted educational leaders to enlarge and improve their educational offerings by developing a more learner- and learning-centered approach to education in order to increase student engagement. An increasing number of educational institutions around the world have implemented undergraduate research because it has been shown to be a high-impact learning activity for students, if designed and implemented properly. This chapter discusses the common elements of successful undergraduate research programs and the common factors that should be considered when designing and implementing such programs. Based on the research from the chapters in this book, the authors propose a generalized undergraduate research model that can be used in a variety of contexts across institutions and grade levels and disciplines.

In Chap. 2 Elizabeth L. Ambos discusses how undergraduate research in the United States has grown in concert with the development of the Council on Undergraduate Research (CUR).

The author discusses the history of CUR and its role as a leader in promoting CUR around the world as well its plan to diversify and expand its role in the future.

In Chap. 3 Angela Brew and Lilia Mantai discuss the development of undergraduate research in Australia over the past decade. The authors explain the challenges in developing a culture of undergraduate research in higher education institutions in Australasia. For instance, they explain that creating the Australasian Conferences of Undergraduate Research was important in establishing undergraduate research in Australasia and how the use of seminars and workshops helped raise awareness of undergraduate research across Australasia and how these efforts led to the establishment of the Australasian Council for Undergraduate Research as a membership organization.

In Chap. 4 Chng Huang Hoon and Wu Siew Mei define undergraduate research as a student-centered inquiry that makes an original academic or creative contribution to a discipline. In this chapter, the authors discuss undergraduate research within the context of National University of Singapore across the disciplines of engineering, science, computing, and arts and social sciences. The authors argue that it is possible to create symbiotic connections between university research and student education by integrating undergraduate research and learning into the curriculum and across disciplines.

In Chap. 5 Amani Elshimi discusses undergraduate research practices at different universities in an attempt to identify an overarching strategy and guiding vision. The author explains that most Egyptian universities focus undergraduate research on problem-solving for environmental and community issues where the student learning experience is shaped within the context of economic development. The author uses the American University in Cairo as a case study and examines the goals, infrastructure, funding, staffing, and outreach of the program. The author highlights the area of university partnerships as well as the alignment of objectives of different units across campus as key factors for undergraduate research success.

In Chap. 6 Rintaro Imafuku provides an introduction to undergraduate research in Japan and discusses the future of undergraduate research in Japan. The author discusses how Japanese higher education institutions have emphasized final- year projects but have increasingly adopted undergraduate research across disciplines as an effective pedagogical approach. The author discusses the challenges associated with implementing undergraduate research such as the absence of a supporting organization. The author recommends developing a community of practice that will allow professors to share their experiences and best practices.

In Chap. 7 Enakshi Sengupta and Patrick Blessinger discuss the benefits and challenges of undergraduate research as presented in the academic literature. The time needed to implement undergraduate research is significant. The authors focus on undergraduate research in India, Malaysia, and the Kurdistan region of Iraq. They discuss the results of survey of faculty members teaching undergraduate students and the faculty experiences are discussed. The findings of the survey suggest that undergraduate research programs in these countries have been limited as a result of financial constraints and lack of support staff.

In Chap. 8 Ana Lucia Manrique and Douglas da Silva Tinti discuss how Brazil has implemented scientific initiation research programs to support undergraduate research. The authors discuss the results of a survey on the implementation of undergraduate research related to the faculty supervisor's projects and lines of research.

In Chap. 9 Roisin Donnelly, Claire McAvinia, and Claire McDonnell discuss the impact of faculty and student learning related to sharing inspirational practices and creating multimedia at a university in Ireland. The authors discuss the increasing demand for professional development, the importance of a peer learning, the development of multimedia artifacts for undergraduate supervision, and the national context for this type of work.

In Chap. 10 Mandla S. Makhanya discusses three areas for integrated research development: niche research themes aligned to national socio-economic priorities, targeted development of inter- and transdisciplinary research, and of higher education institutions. Makhanya discusses the preparation of researchers who will help level the global research playing field and contribute to national development.

In Chap. 11 Isabelle Mirbel and Margarida Romero discuss how a national research ecosystem has been developed to promote excellence in academic research. The authors note that a research-oriented curriculum at the undergraduate level has yet to be considered at a national policy level. The authors also note that, in spite of this, various initiatives exist within higher education institutions to encourage the participation of undergraduate research activities. The authors discuss current initiatives for engaging undergraduate students in research projects, especially those research activities that develop the research competencies that allow undergraduate students to engage in research after graduation.

In Chap. 12 Wolfgang Deicke and Harald A. Mieg discuss how the *Bologna* Reform process and the introduction of a two-cycle BA/MA degree system impacted the German higher education system. The authors highlight how policy initiatives such as the Quality Pact for Teaching (QPT) facilitated new approaches to research-based teaching and learning in German higher education since 2012.

In Chap. 13 Jase Moussa-Inaty discusses how the global trend of students engaging in undergraduate research holds true in the United Arab Emirates (UAE) as well. The author discusses how undergraduate research in the UAE is recognized as an important part of student learning and engagement. The author discusses some of the most recent attempts made to encourage undergraduate research.

In Chap. 14 Rachel Spronken-Smith discusses recent legislation that calls for a close link between research and teaching at New Zealand universities and how this new legislation has put a renewed focus on creating more and higher-quality undergraduate research programs. Yet, the author notes that opportunities for undergraduate students to engage in research are limited. The author also notes that only a fraction of undergraduate programs scaffold the development of research skills throughout the curriculum toward a culminating capstone project. As a result, notes the author, although undergraduate research is pervasive across New Zealand universities, it is not always well supported.

In Chap. 15 Brad Wuetherick discusses how undergraduate research has a strong tradition across Canadian higher education institutions. The author notes that institutions are still challenged to ensure that all undergraduate students are engaged in high-impact research over the course of their university experience. Thus, the author explores the state of both mentored (co-curricular) and curriculum-embedded undergraduate research experiences at universities across Canada.

In Chap. 16 Luísa Soares discusses the undergraduate research experience in Portugal. The author explores whether or not undergraduate students are cognitively and emotionally mature enough to carry out high-quality, scientifically rigorous research. The author argues that it is possible for undergraduate students to carry out rigorous undergraduate research but it must be accompanied with a solid foundation in ethical research principles.

Conclusion

The research findings and case studies presented in this book provide an important knowledge base for those educational professionals thinking about designing, developing, and implementing undergraduate research at their own institution as well as those interested in improving an already existing undergraduate research program. This book not only provides an overview of undergraduate research—its purpose and principles—but it also provides an overview of the current undergraduate research landscape by examining authentic programs and experiences across a diverse set of higher education institutions around the world. As such, for those interested in implementing undergraduate research, this book offers a meaningful guide to that end.

Laguna Woods, CA, USA — Nancy H. Hensel
New York, NY, USA — Patrick Blessinger

Contents

Notes on Contributors

Elizabeth L. Ambos served as the Council on Undergraduate Research's fourth executive officer from 2012 to 2019. From 2006 to 2012, Ambos was assistant vice chancellor for research initiatives and partnerships for the California State University system. Prior to that, she held positions at California State University, Long Beach, including Associate Vice President for research, Graduate Dean, Associate Dean in sciences, and Geology Professor. Ambos received her AB in Geology from Smith College and her MS and PhD from the University of Hawai'i at Mānoa in Marine Geology and Geophysics.

Patrick Blessinger is an adjunct associate professor of education at St. John's University, an educator with the New York State Education Department, and chief research scientist of the International Higher Education Teaching and Learning Association (in consultative status with the United Nations). Blessinger is an educational policy analyst and contributing writer with UNESCO's Inclusive Policy Lab, *University World News*, *The Hechinger Report*, *The Guardian*, and *Higher Education Tomorrow*, among others. Blessinger teaches courses in education, leadership, and research methods and serves on doctoral dissertation committees. Blessinger founded and leads a global network of educators focused on improving teaching and learning and is an expert in inclusion, equity, leadership, policy, democracy, rights, and sustainability. He has received several educational awards, including: Fulbright Senior Scholar to Denmark (Department of State, USA), Governor's Teaching

Fellow (Institute of Higher Education, University of Georgia, USA), and Certified Educator (National Geographic Society, USA).

Angela Brew is Emeritus Professor and previously Professorial Fellow in the Learning and Teaching Centre at Macquarie University, Australia. She is an elected Fellow of the UK's Society for Research into Higher Education (SRHE), a Life Member of the Higher Education Research and Development Society of Australasia (HERDSA), and an Australian Senior National Teaching Fellow.

Chng Huang Hoon is Associate Professor of English Language & Literature and serves concurrently as Associate Provost (Undergraduate Education) and Director (Chua Thian Poh Centre for Community Leadership) at the National University of Singapore. She is a member of the International Society for the Scholarship of Teaching and Learning (ISSOTL) and currently serves on the ISSOTL Board of Directors as co-President Elect and was formerly ISSOTL's Regional Vice President (Asia Pacific). She is active in promoting SoTL, particularly through the regional network called SoTL-Asia, which she established with her NUS colleagues in 2016.

Douglas da Silva Tinti graduated in Mathematics (2006) and Postgraduate Studies in Applied Statistics (2008) from the Methodist University of São Paulo. He holds a master's (2012) and a PhD (2016) in Mathematics Education from the Pontifical Catholic University of São Paulo (PUC-SP). He is Professor, Department of Mathematics Education of the Federal University of Ouro Preto, Minas Gerais, Brazil. His research interests lie in teacher education and mathematics education.

Wolfgang Deicke read Sociology and Politics at Hamburg University and later at the Postgraduate School of Peace Studies in Bradford. Before becoming the coordinator of the bologna.lab at Humboldt-Universität zu Berlin in 2012, he variedly taught Sociology, Politics and the History of European Thought and Society at the (now) University of Northampton, School of Oriental and African Studies in London and Ruskin College, Oxford. In Northampton and Oxford, he gained comprehensive experience in course and curriculum development and the development of curricula for active learning. His current research interests are the development of (student) research competencies and organizational development in higher education.

Roisin Donnelly is Head of Learning Development for the College of Business in the new TU Dublin. This role sees her supporting the College in achieving impact and excellence in teaching and learning through partnership and program provision, promoting enhanced [e]learning strategies and Business Education communities of practice. This builds on her previous positions as Lecturer, Program Chair/Coordinator, Academic Developer for 18 years and as sectoral Project Manager for the initial national implementation of the Professional Development Framework for teachers in Irish higher education. She is a member of the national Professional Development (PD) Advisory Group established to shape how the National Forum for the Enhancement of Teaching and Learning achieves sectoral strategic objectives in relation to PD. She is a fellow of the UK Staff and Educational Development Association (FSEDA), a fellow of the Higher Education Academy (FHEA), and coeditor of the *Irish Journal of Academic Practice*.

Amani Elshimi is Senior Instructor of Rhetoric and Composition, and Founding Director of the Office of Undergraduate Research at the American University in Cairo. She is a devoted advocate of undergraduate student scholarship, and an active practitioner of engaging pedagogies, including community-based learning, senior capstone experiences, and course-based internships. She has over 25 years of experience in higher education teaching, faculty development, curriculum design, assessment, and program administration.

Nancy H. Hensel was the first president of The New American Colleges and Universities, serving from 2004 until 2011. Previously, she served as chief executive officer of the Council on Undergraduate Research in Washington, D.C. for seven years. During her tenure at CUR, she was co-principal investigator for seven National Science Foundation grants to assist faculty and institutions develop undergraduate research programs. She initiated an undergraduate research program at the University of Maine at Presque Isle where she served as president. Prior to her presidency, she was provost at the University of Maine Farmington and professor of education and department chair at the University of Redlands.

Hensel holds a doctorate degree in early childhood education from the University of Georgia, masters' degrees in theater and early childhood education from San Francisco State University and a Bachelor of Arts in theater also from San Francisco State. In 2003, Hensel was inducted into the Maine Women's Hall of Fame for her work in promoting higher education in Maine and supporting the role of women in higher education. She is the author of *Course-Based Undergraduate Research: Educational Equity and High Impact Practice* and several articles and monographs about undergraduate research.

Rintaro Imafuku is an assistant professor of Medical Education Development Center at Gifu University, Japan. He holds an MA (Applied Linguistics) from Monash University and PhD (Education) from the University of Hong Kong. His work focuses on interactional and ethnographic exploring of students' learning in learner-centered inquiry-based educational environments, including PBL, undergraduate research and interprofessional education. He is editing a volume on *Interactional Research into Problem-based Learning* (with Susan Bridges, Purdue University Press) and has guest-edited *Interdisciplinary Journal of Problem-Based Learning*. He is also an editorial board member of several journals, including *BMC Medical Education*.

Mandla S. Makhanya was appointed Principal and Vice Chancellor of the University of South Africa on January 1, 2011 and is a prominent proponent of higher education leadership and advocacy. Makhanya is past President of the International Council for Distance Education (ICDE), Treasurer of the African Council for Distance Education (ACDE), and President of the Higher Education Teaching and Learning Association. Makhanya is a Deputy Chairperson of the South Africa National Commission for UNESCO. He has also been a member of the National Committee of the Memory of the World (MoW). In the 1990s, he served in various leadership roles in the South African Sociological Association, including as Deputy President I 1998, for International Association of Sociology (ISA). Makhanya is on the advisory board of JRODel (*Journal of Research in Open, Distance and e-Learning*).

Ana Lucia Manrique graduated in mathematics, University of São Paulo, Brazil (1987). She holds a master's degree in Mathematics (1994) and a PhD in Education (2003), Pontifical Catholic University of São Paulo (PUC-SP). Manrique obtained a postdoctorate in Education, Pontifical Catholic University of Rio de Janeiro (2008). She is Professor on the Mathematics Education Program, PUC-SP, Brazil and Productivity Researcher at CNPq (PQ2). She has coordinated projects financed by Brazilian development agencies (Capes, CNPq, and Fapesp). Her research interests lie in inclusive education, teacher education, and mathematics education.

Lilia Mantai is Academic Lead, Course Enhancement at The University of Sydney Business School. She was awarded a PhD for research on researcher development of doctoral students in 2017 at Macquarie University. She is a senior fellow of the Advance HE/Higher Education Academy.

Claire McAvinia is an academic developer at the Learning, Teaching and Technology Centre (LTTC) in TU Dublin. She is currently LTTC Programs Chair and lectures on the Centre's accredited postgraduate programs and modules in learning and teaching, and e-learning. Her interests are in academic development, digital literacies, curriculum design, and open education. Claire is engaged in supervision of students in master and doctoral programs. She has published on a range of topics in e-learning and academic development, and in 2016 completed a book based on her doctoral research, *Online Learning and its Users: Lessons for Higher Education*. Claire is Fellow of the UK Staff and Educational Development Association (SEDA) and the UK Higher Education Academy. Together with Dr. Roisin Donnelly and Dr. Kevin O'Rourke, she coedits the *Irish Journal of Academic Practice* (IJAP).

Claire McDonnell is Assistant Head at the School of Chemical and Pharmaceutical Sciences in TU Dublin. Her interests include facilitating learner transition to higher education and the application of technology to support student learning and collaboration. She has implemented several approaches to embed research and enquiry skills in the curriculum, including context- and problem-based learning and community-engaged learning. She spent a 3-year secondment with the TU Dublin Learning,

Teaching and Technology Centre from 2013 to 2016 where she was program coordinator for MA in Higher Education. Claire is a founding member of the Chemistry Education Research Team at TU Dublin, which won a DELTA award from the Irish National Forum for the Enhancement of Teaching and Learning in 2018.

Harald A. Mieg is (Honorary) Professor of Metropolitan Studies at the Humboldt-Universität zu Berlin and also affiliated to the Swiss Federal Institute of Technology, Zurich. He was the initiator and coordinator of a German national research project on undergraduate research in Germany ("ForschenLernen," during 2014–2018; 15 universities were involved). In this context, he coordinated the edition of a series of books on undergraduate research and inquiry-based learning in Germany.

Isabelle Mirbel is an associate professor in computer sciences. Her research activities are in the fields of information systems engineering, method engineering and requirements engineering. From 2012 to 2018, she was a vice-dean of the Faculty of Sciences in charge of professional integration and relations with companies and then in charge of pedagogy. For the past two years, she has been coordinating activities to implement competency-based education within the Faculties of Université Côte d'Azur. She is also in charge of the "Sciences à la carte" project, in collaboration with the digital thematic university, Unisciel, which aims to develop online digital resources and more precisely self-positioning tests to facilitate the orientation of undergraduate students.

Jase Moussa-Inaty is Associate Professor of Educational Psychology at Zayed University. She is actively engaged in conducting research on various topics such as cognitive load, metacognition, enhancing thinking skills, blended learning, motivation, self-efficacy, science anxiety, parental involvement, and effective teaching and learning processes. She has been heavily involved in undergraduate research and was one of the founding researchers to establish the Undergraduate Research and Creative Projects symposium at Zayed University.

Margarida Romero is a research director of the Laboratoire d'Innovation et Numerique pour l'Éducation (LINE), a research lab in the field of Technology Enhanced Learning (TEL, full-time professor at Université Côte d'Azur (France) and an associate professor at Université Laval (Canada). Her research is oriented toward the inclusive, humanistic, and creative uses of technologies (co-design, game design, and robotics) for the development of creativity, problem solving, collaboration, and computational thinking.

Enakshi Sengupta is Associate Director of the International Higher Education Teaching and Learning Association (HETL) and is responsible for the advancement of HETL in Asia, Middle East, and Africa. Sengupta is also the Director of the Center for Advanced Research in Education (CARE), Associate Series Editor of the book series, *Innovations in Higher Education Teaching and Learning*, Emerald Group Publishing. She is Managing Editor of the *Journal of Applied Research in Higher Education*, Emerald Publishing, and serves as Vice Chair of the Editorial Advisory Board of the *Innovations in Higher Education Teaching and Learning* book series, Emerald Publishing. Sengupta is Senior Manager of the Research, Methodology, and Statistics in the Social Sciences forums on LinkedIn and Facebook responsible for managing all aspects of those forums. Sengupta holds a PhD from the University of Nottingham in research in higher education, prior to which she completed her MBA with merit from the University of Nottingham and master's degree in English Literature from the Calcutta University, India. Sengupta has previously held leadership positions in higher education institutions.

Luisa Soares is a researcher at Larsys – Interactive Technologies Institute. Clinical psychologist, coordinator of psychology service at Madeira University and Course Director of Psychology course 1st cycle, she teaches Clinical Psychology, Neuropsychology and Psychopathology of infancy and adolescence.

Rachel Spronken-Smith is a professor in Higher Education and Geography. She holds a PhD in Geography (UBC, Canada) and PGDip. in Tertiary Teaching (Otago, NZ). After lecturing in Geography at Canterbury University, she moved into Higher Education at the University of Otago, NZ. Rachel has led national projects on inquiry-based learning and graduate attributes. She has won teaching awards at both Canterbury and Otago, and a national tertiary teaching award in 2015. In 2016, she won the TERNZ-HERDSA medal for Sustained Contribution to the Research Environment in NZ, and gained a Fulbright Scholar Award for doctoral education research in 2018.

Wu Siew Mei is an associate professor of Applied Linguistics and Director at the Centre for English Language Communication, National University of Singapore. She currently serves on the Editorial Board for the *Asian Journal for the Scholarship of Teaching and Learning*. She is a key leader in the regional network called SoTL-Asia and has published with *Higher Education Research and Development Society of South Australasia* (HERDSA).

Brad Wuetherick is Executive Director, Learning and Teaching in the Office of the Provost and VP Academic and Centre for Learning and Teaching at Dalhousie University, Atlantic Canada's preeminent medical-doctoral research university. He has been actively engaged in researching, and supporting the implementation of, undergraduate research since 2003, while working at three different medical-doctoral research universities across Canada.

List of Figures

List of Tables

CHAPTER 1

Undergraduate Research as a High-Impact Educational Practice

Patrick Blessinger and Nancy H. Hensel

Introduction

Over the past several decades, the demand for higher education around the world has increased substantially (UNESCO 2018). Several factors have fueled the growing demand. As higher-education institutions expand their services to accommodate the increased demand for education, they have also put greater emphasis on academic engagement and quality to increase student retention and graduation rates. UNESCO (2018) estimates that by 2040 nearly 600 million students will be enrolled in colleges and universities around the planet, up from 216 million students in the year 2016.

This statistic is striking when one considers that the global college-age population will reach 800 million in 2040. Much of the growth in

P. Blessinger (✉)
International Higher Education Teaching and Learning Association,
New York, NY, USA
e-mail: patrickblessinger@gmail.com

N. H. Hensel
Laguna Woods, CA, USA
e-mail: hensel.nancy@gmail.com

N. H. Hensel, P. Blessinger (eds.), *International Perspectives on Undergraduate Research*,
https://doi.org/10.1007/978-3-030-53559-9_1

college-age people will come from African countries. Also reflected in these numbers are millions of additional students who enroll in free courses through open education platforms like MIT's OpenCourseWare as well as nonformal learning platforms like Khan Academy. It is fair to say that in many countries higher education has reached a universal status, as defined by Martin Trow (UNESCO 2018). Since higher education is viewed by many as a critical factor in attaining employability and social mobility through knowledge acquisition and skill development, it comes as no surprise that many people are enrolling in colleges and universities, and other learning programs, in record numbers.

This increased demand in higher education has also created increased competition for students, which, in turn, has put greater pressure on institutions to improve their services. One way they are doing this is by implementing undergraduate research programs, mainly because it is seen as one of the most beneficial high-impact learning activities in existence (Kuh and Hu 2001; Kuh 2008). Campuses have acknowledged the documented benefits of undergraduate research and expanded opportunities for student engagement in research. It is vital to understand deeply the impact of undergraduate research so that context-sensitive ways of improving UGR may be found, whether that be small tweaks, major adaptations or implementations of whole new models (Komarraju et al. 2010; Lopatto 2006, 2007; Webber et al. 2012).

The benefits of undergraduate research include but are not limited to: improved faculty teaching performance, increased faculty–student collaboration/mentoring, increased student engagement, improved teamwork skills, increased academic achievement, higher-order thinking and inquiry skills, improved perseverance in problem-solving, and increased self-confidence. These benefits, in turn, help improve student persistence and retention.

Intellectual, psychological, and social characteristics are also crucial in preparing students for graduate studies and professional employment. In some cases, involvement in undergraduate research helps students to reevaluate and fine-tune their career choices, especially for those who may still be unsure about what graduate program or career to go into (Kuh 2008; Gentile et al. 2017).

Undergraduate research is identified as a high-impact learning practice and, as such, it is linked to improved student achievement and institutional advancement. Undergraduate research can take different forms depending on how it is utilized. With respect to scope of integration, it can be used as a one-off extracurricular activity or it can be a one-off curricular activity where it is integrated into a single course as a learning activity or it can be integrated across a set of related courses as part of a broader program or departmental undergraduate research effort where the research continues across semesters (Lopatto 2009; Gentile et al. 2017).

With respect to level of research, it can be used as one of several different types of learning activities a student must perform, say with lower or equal grade weighting, or it can be the main learning activity that is weighted relatively heavily. The level of the research refers to the intensity of the research performed. The level of research is on a spectrum of intensity from low level to medium level to high level. A low-level (i.e., low intensity) research activity may only involve collecting and analyzing secondary data. In contrast, a high-intensity research activity involves collecting and analyzing primary data. The level of research performed is often a function of several factors, including grade and course level, and the degree of collaboration with the faculty member or mentor.

With respect to type of research, it can be used with any of the main learning domains (arts, humanities, science), and thus, the type of research will depend on the specifics of the domain, the discipline, the field, and the particular course. For instance, STEM fields use the scientific method as the main problem-solving approach and they rely on experimental methods to collect and analyze empirical data. In nonscientific fields such as the arts and humanities, they also use evidence (data) to carry out research but the data is often qualitative, collected and analyzed by nonexperimental methods. The common denominator across the domains is the use of the research process to collect evidence to answer research questions (Blessinger 2017; Hensel 2018). Appendix A provides a high-level overview of the research process.

The faculty member is in the best position to determine the scope, level, and type of research to be conducted by students, how best to scaffold the research process, and the nature of the research question to be answered. Regardless of the scope, level, and type of research conducted, the ultimate objective of undergraduate research is for the students to make, to one degree or another, an *original* contribution to the field of study related to the course. Although students may work independently to carry out the research, they typically collaborate with a faculty member. The degree of collaboration will depend on the scope, level, and type of research, among other factors (Kinkead and Blockus 2012; Gentile et al. 2017).

The Purpose and Value of Undergraduate Research

Apart from the benefits of undergraduate research, it is essential to understand the underlying mechanisms that create the value proposition of undergraduate research, and that defines its purpose. Research is the process of conducting a systematic inquiry to produce original knowledge. Research involves the systematic analysis of data in order to answer research the question(s). Thus, the purpose of research is to generate new knowledge, which involves collecting and analyzing data, both secondary and primary. The type of data collected and the way the data is analyzed will depend on the research methodology used, which in turn, will depend on such factors as the research question(s) and the research objectives as well as the knowledge domain, discipline, and field. Research is an inquiry-based learning activity, so research begins and ends with the research question and objectives.

Embedding undergraduate research into a course is not the only way to implement undergraduate research. The apprenticeship model—one professor and a few students—is another way to implement undergraduate research. Other ways to implement undergraduate research might include integrating it with study abroad experiences, internships, service-learning, and student learning communities, among others. Undergraduate research can also be integrated into extracurricular activities. Any activity that would lend itself to the research process could be a viable candidate for undergraduate research.

Although collecting primary data is often considered the heart of the undergraduate research experience, collecting and analyzing secondary data is also essential. Since research is a process of inquiry that can be used in any discipline or field, it lends itself to a wide variety of ways to integrate it into courses and programs. On one end of the research spectrum, it could be limited to researching secondary data, such as reading and analyzing data from existing literature. This type of research might be considered low-intensity or low-end research. In contrast, research involving the collecting and analyzing of original, primary data could be considered high-intensity or high-end research. High-intensity research incorporates more intense higher-order thinking, the use of more structured research methodologies, and collaboration with faculty or other mentors.

High-intensity research is inclusive of secondary data research. Data that comes from existing sources (e.g., journals, books, reports) is vital as it allows the researcher to understand what research has already been conducted, and it allows the researcher to identify any knowledge gaps in the field. Conducting a review of the existing literature is a prerequisite part of conducting primary research. Conducting secondary research is important research in itself. Developing expertise in conducting secondary research is important regardless of what domain or discipline or field one is in. Conducting secondary research allows students to increase their knowledge on the subject prior to conducting primary research.

Undergraduate research is a data-driven process. As such, the quality of the data is of paramount importance. By conducting research, students learn to understand the importance of data quality and the methods and procedures that can be used to help ensure data quality. Several characteristics define data quality, including the accuracy and completeness of the data, as well as timeliness and relevance of the data. Appropriate procedures must be followed in order to maintain the high quality of the data. The specific procedures and protocols used in undergraduate research will depend on the knowledge domain, discipline, and field, as well as the particular research methodology used. Ultimately, however, the quality of the data is determined by how well it allows or aids the researcher to answer the research questions and meet the research objectives.

Research ethics and professional standards are other areas of concern for undergraduate researchers. Research involving the participation of humans (especially children) or the use of animals requires adherence to strict research ethics and protocols and often requires the approval of a research ethics board or an institutional review board. With human subjects' research, the researcher must be trained in research ethics (e.g., principles of informed context and do no harm as well as issues of confidentiality). Human subjects' research also requires direct supervision from an approved faculty member.

Thus, given the legal and ethical issues involved in human subjects' research, undergraduate research is often confined to research that does not utilize human subjects for the purpose of data collection. Even with research involving animals, proper training must be given to students so that they adhere to established protocols, procedures, and ethics. Engaging undergraduate students in research allows students to be inculcated in the research process and initiated into the research community as well as involving them in those areas that impact research (Blessinger 2017; Hensel 2018).

The Undergraduate Research Experience

In addition to increased student engagement and increased academic achievement, undergraduate research also helps to improve students' work habits, communication skills, and career choice clarification. As an authentic form of experiential learning, undergraduate research can take the form of an apprenticeship wherein they learn how their field of study operates through hands-on training. Not only do students acquire disciplinary knowledge, but they also learn the research skills specific to their field. These skills involve all those elements common to the research process: research strategy and design, data collection and analysis, and communication of results. Students also have the opportunity to publish their results, where they become part of the community of practice (Kuh 2008; Lopatto 2010; Gentile et al. 2017).

Auchincloss et al. (2014) identified five dimensions of the undergraduate research experience from a faculty point of view: use of scientific practices, discovery, broadly relevant or important work, collaboration, and iteration. Based on these dimensions, Gentile, Brenner, and Stephens (2017, pp. 34–35) have identified the following core characteristics as important to the undergraduate research experience in the STEM disciplines:

- Emphasis on engaging students in research practices and arguing from evidence
- Emphasis on generating new data and replicating preliminary results
- Emphasis on significant and relevant problems of interest to STEM researchers
- Emphasis on collaboration and teamwork
- Emphasis on research questions, experimental design, and data collection
- Emphasis on mastering research techniques
- Emphasis on engaging students in problem-solving reflection
- Emphasis on communication of results
- Emphasis on mentorship and increasing ownership of the research project over time

The high-level process is also similar for non-STEM fields in that the research process is an evidence-driven, inquiry-based, problem-solving process. Fundamentally, undergraduate research is an investigation into a research question that the researcher wants to answer. Therefore, the particulars of the problem-solving process are specific to the domain and the discipline and the field of study. So, the research topic and the research question(s) will also be specific to the domain, discipline, and field. Research in the arts or the humanities, for instance, is just as rigorous and beneficial and important as research in the natural and social sciences. Domains are inherently different from another, not inherently better than another (Blessinger 2017; Hensel 2018).

In undergraduate research, students typically work with mentors to help guide them through the process. Depending on the type of research being undertaken, the mentor may be a faculty member or a graduate student or a postdoc researcher or even an industry professional researcher. Mentorship and collaboration are key to a successful undergraduate experience as it has been positively correlated with increased academic achievement in students.

In addition, high-quality mentoring is key to improving a student's continued education in the research field. Finally, research allows students to learn more than what is in the curriculum (e.g., textbooks) and more than what is provided from lectures and classroom discussions. Thus, undergraduate research is a high-impact learning activity because it allows students to operate at the top end of Bloom's Taxonomy (i.e., higher-order thinking) for sustained periods of time (Kuh 2008; Lopatto 2009). It also builds up mental stamina and critical thinking skills.

Types of Undergraduate Research

There are many different types of undergraduate research. As mentioned before, the type of research conducted depends on several factors, and that combination of factors will ultimately drive the experience for the student. For instance, Gentile, Brenner, and Stephens (2017, p. 35) note that, "developing technical skills and knowledge is often a focus in early research learning experiences, while opportunities to learn how to deal with failure and develop resiliency tend to emerge as students get more deeply involved in a research project."

In addition, Brownell and Kloser (2015) note that a course limited solely to secondary data collection is unlikely to educate students on how to actually conduct research. Further, a course in research methods may teach the technicalities of research but may not engage students in actually doing research competently. The key to developing higher-order research skills is to learn by doing in order to internalize the full depth and breadth of real-world research. Undergraduate research may go by different names, depending on the institution, the domain, the discipline, the field of study, etc. Gentile, Brenner, and Stephens (2017) have identified several different types of undergraduate research that may go by different names:

- Individual faculty research (student as apprentice)
- Cumulative capstone courses and senior theses
- Internships and co-op experiences
- Study abroad and international programs
- Project-based and problem-based courses and programs
- Community-based and field-based programs

There are many ways to integrate undergraduate research into the curriculum (or outside the curriculum for that matter) and several models on how to create a meaningful undergraduate research experience for students. Although much has been written about these undergraduate models (Kierniesky 2005; Kortz and van der Hoeven Kraft 2016; Pukkila et al. 2007; Reinen et al. 2007; Rueckert 2007; Temple et al. 2010), much can still be learned by examining the current state of undergraduate research today.

Modeling the Undergraduate Research Experience

The undergraduate research experience is multifaceted and varies by discipline, mentoring type, duration, complexity, etc. Depending on how these factors come together for the student and how the faculty member scaffolds the research activities for the student, the outcomes of undergraduate research will therefore vary with respect to its impact on one's future studies, to one's career choice, etc. Given the diverse nature of undergraduate research, it is difficult to develop a single model that applies in all situations. Nonetheless, based on a review of the literature and the chapters in this volume, Fig. 1.1 attempts to provide a high-level, process-oriented and product-oriented view of undergraduate research to serve as a general guide in designing, developing, and implementing undergraduate research.

This model does not offer a cookbook approach to undergraduate research but rather a high-level guide to steer faculty members, mentors, and others in the right direction. As the model shows, the undergraduate

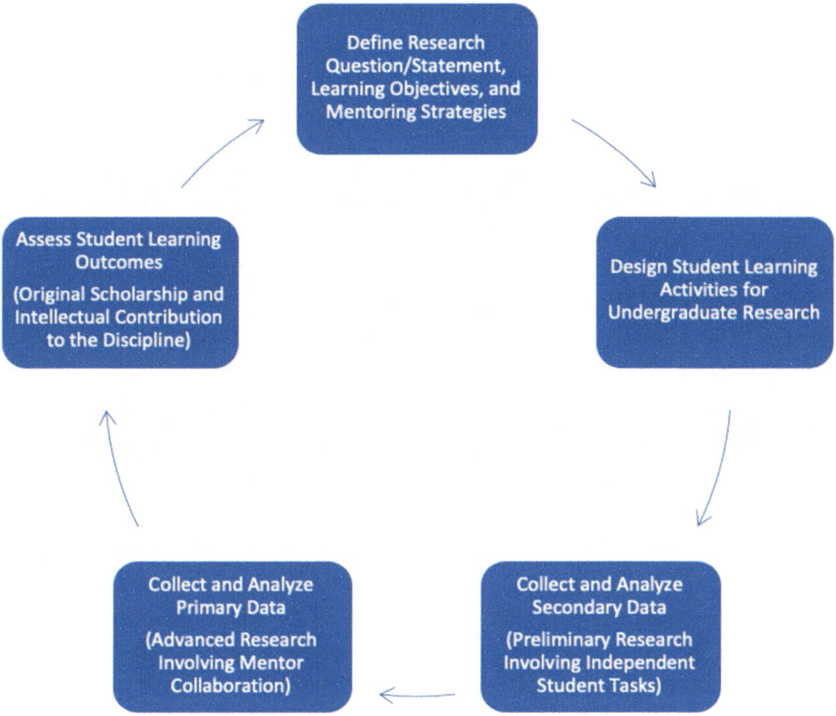

Fig. 1.1 Undergraduate research model

research experience should be embedded within the learning objectives of the course the student is taking or within the learning objectives of the program the student is involved in, which itself is likely to be embedded in a domain, discipline, and field of study. Thus, the learning objectives that the student is expected to achieve drives the whole process and drives the goals of the undergraduate research experience. The instructor or program manager should carefully reflect on and define those specific learning objectives in order to scaffold the research activities appropriately. The model also serves as a guide in the course or program design process involving undergraduate research.

Define Course Learning Objectives and Teaching Strategies In order for course learning objectives to be meaningful, they must be coherent, specific, and measurable. In a learner-centered environment, they must also be written from the learners' point of view and with the students in mind. To effectively design a course or learning program, each learning objective should be in response to the following statement: "by the conclusion of the course, students will be able to…."

Each learning objective should then conclude with a specific, actionable outcome the students are to attain, with respect to knowledge and skills. The words and phrases used in crafting the learning objectives should reflect the level of thinking and the mastery of knowledge and skills (e.g., using Bloom's Taxonomy) that students are expected to achieve. In addition, the assessments given (both formative and summative) should be consistent with the learning objectives and in alignment with the undergraduate research expected to perform.

For example, a learning objective that is limited to secondary data sources, say for an introductory course or a lower-level course or, for instance, a course in world history, might look like: Explain the effect of the Great Depression on the rise of fascism around the world by writing an academic essay that analyzes and evaluates why and how Mussolini was able to seize power in Italy. Alternatively, the statement could be phrased in the form of a research question: How did the Great Depression affect the rise of fascism around the world and why and how was Mussolini able to seize power in Italy? Another learning objective for this course might look like: Explain the effect of immigration on German culture in the twenty-first century by writing an academic essay that analyzes and evaluates why and how German immigration policies led to a massive influx of immigrants. In both objectives, students are expected to engage in higher-order thinking skills.

Primary data is any type of data the student is expected to collect himself/herself. There are several ways to collect primary data, such as observations, interviews, and surveys. If students have not had a course in research methods, the course designer must design learning objectives based on the students' background knowledge and level of experience in the subject matter and based on their current research knowledge and skills. A learning objective that includes collecting and analyzing primary data also involves higher-order thinking skills but the research process is at a higher level of intensity. For example, a learning objective that includes collecting primary data, say for a course in public health, might look like: Explain the effects of New York City's proposed ban on soft drink size by surveying a random sample of New York City residents and then discuss your findings in the form of a research report.

Design Student Learning Activities for Undergraduate Research After crafting the learning objectives, the course designer must design the student learning activities that the student must perform to fulfill the objective. The learning activity is a more detailed explanation of the learning objective. If the objective requires only secondary research, then the course designer must explain the criteria and the scope of the activity and provide the student with guidance about how to fulfill the objective. In order to maximize the learning for students, learning activities must be carefully designed by the course designer (e.g., faculty, mentor, program manager).

Collect and Analyze Secondary Data Even if the learning activity involves collecting primary data, research projects should also involve collecting and analyzing secondary data because such projects involve conducting preliminary research in order to discover the knowledge that currently exists on the topic. Examining secondary data on a topic is important because it also helps the researcher understand what the key issues are for a given topic, and it helps identify the knowledge for a given topic. In short, looking at secondary data better prepares the researcher to properly carry out the primary data collection.

Some research projects require the collection of primary data because there may be little written about the topic in secondary sources or because primary data is needed to gain a deeper understanding of the topic and to fulfill the learning objectives. To become an expert in research ethics and research methods, it is important to know how to use both secondary and primary data sources and how to integrate the two different sources into the final research product.

Collect and Analyze Primary Data In primary research, certain established strategies are often used. These strategies are typically classified as quantitative strategies or qualitative strategies or mixed strategies. The strategy used depends on the research questions and learning objectives one is trying to achieve. In quantitative research, for instance, typical strategies used include nonexperimental and experimental strategies, as well as the meta-analysis strategy. In qualitative research, for instance, typical strategies used include grounded theory, ethnography, phenomenology, narrative inquiry, and historical research. Mixed research includes case studies, action research, and assessment research.

Whereas the research strategy defines the high-level approach to be used and defines the focus of the research, the research analysis part of the project defines the data collection techniques used and the data analysis procedures used. For quantitative data (numeric data from samples), typical techniques used include questionnaires, interviews, observations, and tests. For qualitative data (text, images, videos), typical techniques used include questionnaires, interviews, focus groups, and observations. These methods are commonly used in the humanities and social sciences. The arts will have their own methods as will the natural sciences, which rely primarily on experimental methods.

Assess Student Learning Outcomes An assessment helps determine how well the student has mastered the topic (i.e., achieved the learning objective) in those areas that are being assessed (i.e., knowledge, skills, etc.) with respect to appropriate learning standards and competencies. No matter what type of assessment is used, the assessment should reflect the learning objectives for the course and the learning activity. Bloom's Taxonomy, for instance, is a helpful guide because it informs one about the level of learning taking place. The assessment must be in alignment with the learning objective(s) and activities. The assessment must accurately and completely assess the level, scope, and type of learning that occurred. Thus, the assessment may take the form of a test, an essay, a report, or a project, among others, and it should include a scoring rubric so that both the students and the assessor know what criteria will be used to grade the assessment.

In addition, prior to designing an undergraduate research learning activity, it is helpful to first think about the relevant questions involved in doing so. Though not a comprehensive list, when thinking about whether or not to integrate undergraduate research into the curriculum or into a specific course, one should ask the following questions:

> What are the benefits and advantages of undergraduate research?
> How will undergraduate research enhance the course?

What are the likely costs and benefits for the students?
What are the likely challenges and prerequisites for the students?
What are the responsibilities and expectations for the students?
What are the responsibilities and expectations for the faculty?
What specific undergraduate research (scope, level, type, etc.) will the students undertake?
How are the research topic, research questions, and research objectives determined?

These questions will allow the course designer to reflect more deeply on how best to design the undergraduate research experience to increase student engagement and student success and thereby improve student retention and graduation rates.

Conclusion

Undergraduate research is becoming more important around the world as colleges and universities strive to improve their undergraduate experience. More institutions around the world are implementing it into their student experience because it has been shown to be a high-impact educational practice. Depending on the institution and the country and other factors, different models of undergraduate research are being used. As undergraduate research becomes more international, as illustrated through the chapters in this volume, institutions are learning from each other about the best way to implement it within their own context.

Appendix A

Educational Research Methodology Framework: Patrick Blessinger (2017)

Research phase	*Attributes/Key questions:*
Research question Review the academic literature (lit review) on your research topic to better understand the existing knowledge base related to it	**What do you want to investigate and why ?** All educational research attempts to better understand (meaning-making)a particular phenomenon and the nature of relationship among variables through a systematic inquiry-based analysis and interpretation of data

(*continued*)

(continued)

Research phase	*Attributes/Key questions:*
Define the specific research topic	The topic area of the research project
Define the specific research aim	The purpose and scope of the research project
Define the specific research question(s) you want to answer	The specific research question(s) you want to answer
Define the specific research objective(s)	The question put in the form of specific research objectives
Develop a written research plan proposal	Your *research plan* should describe how you propose so conduct the research; it includes the following sections: introduction, purpose, literature review, research question(s), research strategy, research methods (participants/sample, data collection instruments, data analysis procedures), and conclusion
In a research study, everything begins and ends with the research question(s) you want to answer	
Research perspective researcher(s) *worldview* & assumptions)	**What worldview guides your investigation of the research question(s)?**
Philosophies (view on the nature of reality and knowledge)	
Positivism (objective reality, socially independent)	Interprets reality mainly via value-free, scientific test data
Realism (objective reality, socially dependent)	Interprets reality mainly via senses and social conditioning
Interpretivism (subjective reality, socially constructed)	Interprets reality mainly via symbols/meaning/values/roles
Pragmatism (multiple realities/views acceptable)	Best research design depends mainly on the nature of research question
Approaches problem-solving seasoning)	
Deductive (mainly a positivist approach)	Conclusion deduced from empirical facts; typically tests hypothesis/theory
Inductive (mainly an interpretivist approach)	Conclusion inferred from empirical facts, typically build hypothesis theory
Research design (research strategy used)	**How will you answer the research question(s)?**

(*continued*)

(continued)

Research phase	*Attributes/Key questions:*
Strategic (research design *strategy* for collecting and analyzing data; the strategy most appropriate depends on research questions)	Strategy used will determine what type(s) of data will be collected:(Quant: random or nonrandom sampling, Qual: purposive sampling)
Quantitative (uses sampling and *statistics* with logic & theory) Survey, correlational, causal-comparative, experimental (single subject, quasi, true: to test null hypothesis), and meta-analysis (research about previous research)	Focus is mainly on controlled context to *test* hypotheses. Quantitative designs operate on continuous from descriptive to relational to predictive to cause effect using descriptive statistics and inferential statistics
Qualitative (uses sampling and *coding* with logic & theory)	
Grounded theory (emerged from sociology)	Focus is real life context to build hypothesis or theory
Ethnography (emerged from anthropology)	Focus is real life context and personal stories via their cultural context
Phenomenology (emerged from philosophy & psychology)	Focus is real life context to explain personal meaning of person/group
Narrative inquiry (multidisciplinary)	Focus is real life context and meaning from stories told by the individual
Historical research (multidisciplinary)	Focus is die examination of a past event, activity, person, subject, place, etc
Quantitative or Qualitative or Mixed	
Case study research (multidisciplinary)	Focus is real life context (defined by unit of analysis, not by methodology)
Action research (emerged from organizational behavior)	Focus is organizational context to create change (research by actors for actors)
Evaluation research (multidisciplinary)	Focus is on the merit of a program, policy, process, need, activity, etc
Mixed method research (multidisciplinary)	Combine quantitative and qualitative method: include exploratory research
Methods (techniques and procedures—see below)	(what) and explanatory research (how, why) and the triangulation of methods
Mono (1 data collect technique and l analysis procedure)	
Multiple (>1 data collect technique and analysis procedures)	

(*continued*)

(continued)

Research phase	*Attributes/Key questions:*
Tuneframe	
Cross-sectional	Study a particular phenomenon at a particular time
Longitudinal	Study change and development over a period of time
Research analysis (data methods used)	**Who (sample), what (data), when, where, how (techniques/procedures), and why will you collect/analyses data relevant to the research question(s)?**
Techniques (data *collection* techniques: participates/ instruments)	
Qualitative Data Surveys (questionnaires, interviews, observation), Tests (scores), Documents/Records/ Artifacts Quantitative Data Surveys (questionnaires, interviews/focus groups, observations), Documents/ Records/Artifacts	Collects mainly numeric data from sample for *statistical* analysis. Random sampling: random, simple, stratified, cluster, systematic, nonrandom sampling: convenience, purposive, quota). Random sampling with controls is preferred. Collects mainly nonnumeric data from sample for *conceptual* analysis (which purposive sampling method to use—intensity, homogenous, criterion, snowball, or random purposive—depends on the nature of the study)
Procedures (data *analysis* procedure)	
Quantitative Data (data analyzed *statistically* by researcher(s) using statistics, tables, charts)	Analyzes mainly *numeric and categorical data*. Analyzes independent and dependent variables across different scales: nominal (categorical), ordinal (ranked), and interval/ratio. Tests: descriptive stats (frequencies, percentages, X, SD) & inferential stats (correlations, regressive, t, ANOVA, Chi-square, etc.)
Qualitative Data (data analyzed *conceptually* by researcher(s) using codes, categories, themes)	Analyzes mainly *non-numeric data* (words, images, videos) that are usually coded through thematic analysis, then translated into overarching themes
Research conclusions (researches(s) *interpretations* of the data)	**What have you learned from your research?**
Explain your findings (results of the data analysis)	What results did you analysis reveal? Are they reliable and valid?
Discussion (researches(s) reflection on the findings)	How did you interpret the results and why (e.g., relative to existing theory)?
Draw your main conclusions (key points)	*your answer(s) to your research question(s)*
Discuss the implications for future research	Based on your conclusions, what are the implication for future research?

Sources: Frankel et al. (2009) *How to design and evaluate research*... Gay et al. (2009) *Educational research*... Saunders et al. (2009) *Research methods*...

References

Auchincloss, L. C., Laursen, S. L., Branchaw, J. L., Eagan, K., Graham, M., Hanauer, D. I., Lawrie, G., McLinn, C. M., Pelaez, N., Rowland, S., Towns, M., Trautmann, N. M., Varma-Nelson, P., Weston, T. J., & Dolan, E. L. (2014). Assessment of course-based undergraduate research experiences: A meeting report. *CBE–Life Sciences Education, 13*(1), 29–40.

Blessinger, P. (2017). Transforming learning through student research. *University World News*. Retrieved from https://www.universityworldnews.com/post.php?story=20170530085248669

Brownell, S. E., & Kloser, M. J. (2015). Toward a conceptual framework for measuring the effectiveness of course-based undergraduate research experiences in undergraduate biology. *Studies in Higher Education, 40*(3), 525–544.

Gentile, J., Brenner, K., & Stephens, A. (2017). *Undergraduate research experiences for STEM students: Successes, challenges, and opportunities*. Washington, DC: The National Academies Press.

Hensel, N. (2018). *Course-based undergraduate research: Educational equity and high- impact practice*. Sterling: Stylus Publishing.

Kierniesky, N. C. (2005). Undergraduate research in small psychology departments: Two decades later. *Teaching of Psychology, 32*(2), 84–90.

Kinkead, J., & Blockus, L. (2012). *Undergraduate research offices & programs: Models and practices*. Washington, DC: Council on Undergraduate Research.

Komarraju, M., Musulkin, S., & Bhattachary, G. (2010). Role of student-faculty interaction in developing college students' academic self-concept, motivation, and achievement. *Journal of College Student Development, 51*(3), 332–342.

Kortz, K. M., & van der Hoeven Kraft, K. J. (2016). Geoscience education research project: Student benefits and effective design of a course-based undergraduate research experience. *Journal of Geoscience Education, 64*(1), 24–36.

Kuh, G. (2008). *High-impact educational practices: What they are, who has access to them, and why they matter*. Washington, DC: Association of American Colleges and Universities.

Kuh, G. D., & Hu, S. (2001). The effects of student-faculty interaction in the 1990s. *Review of Higher Education, 24*, 309–332.

Lopatto, D. (2006). Undergraduate research as a catalyst for liberal learning. *Peer Review, 8*(1), 22–25.

Lopatto, D. (2007). Undergraduate research experiences support science career decisions and active learning. *CBE-Life Sciences Education, 6*, 297–305.

Lopatto, D. (2009). *Science in solution: The impact of undergraduate research on student learning*. Tucson: The Research Corporation for Science Advancement.

Lopatto, D. (2010). Undergraduate research as a high impact student experience. *Peer Review, 12*(2), 27–30.

Pukkila, P., DeCosmo, J., Swick, D. C., & Arnold, M. S. (2007). How to engage in collaborative curriculum design to foster undergraduate inquiry and research in all disciplines. In K. K. Karukstis & T. E. Elgren (Eds.), *Developing and sustaining a research-supportive curriculum: A compendium of successful practices* (pp. 341–357). Washington, DC: Council on Undergraduate Research.

Reinen, L., Grasfils, E., Gaines, R., & Hazlett, R. (2007). Integrating research into a small geology department's curriculum. In K. K. Karukstis & T. E. Elgren (Eds.), *Developing and sustaining a research-supportive curriculum: A compendium of successful practices* (pp. 331–339). Washington, DC: Council on Undergraduate Research.

Rueckert, L. (2007). Flexible curricular structures to provide time for research within the classroom. In K. K. Karukstis & T. E. Elgren (Eds.), *Developing and sustaining a research-supportive curriculum: A compendium of successful practices* (pp. 285–294). Washington, DC: Council on Undergraduate Research.

Temple, L., Sibley, T., & Orr, A. J. (2010). *How to mentor undergraduate research*. Washington, DC: Council on Undergraduate Research.

UNESCO Institute for Statistics (UIS). (2018). *Education stats database*. Retrieved from http://uis.unesco.org/

Webber, K., Nelson Laird, T. F., & BrckaLorenz, A. (2012, June 4). *Student and faculty member engagement in undergraduate research*. Paper presented at the Association for institutional research 2012 annual conference, New Orleans, LA.

CHAPTER 2

Undergraduate Research in the United States: Diversity, Growth, and Challenges

Elizabeth L. Ambos

History of Undergraduate Research in the United States and CUR's Origins and Development

The theory and practice of undergraduate research (UR) in the United States (US) emerged from a number of overlapping intellectual, social, and education traditions. The goal of this chapter is to represent US UR's history, current practices, and future directions, while making connections to the wider international UR movement. To the latter point, I would begin by citing the seminal influence of eighteenth-century German polymaths (and brothers) Alexander and Wilhelm von Humboldt on US UR origins. Alexander von Humboldt pioneered many of the fields of natural science that engage current-day researchers and had a remarkable facility for communicating and globalizing scientific investigations (e.g. Wulf 2015). Wilhelm von Humboldt envisioned, designed, and helped to implement Prussian university educational systems conjoining original

E. L. Ambos (✉)
Council on Undergraduate Research, Washington, DC, USA
e-mail: beth@theyer.com

N. H. Hensel, P. Blessinger (eds.), *International Perspectives on Undergraduate Research*,
https://doi.org/10.1007/978-3-030-53559-9_2

research and instruction, a model that many German (and US) institutions strive to encompass today (e.g. Mieg 2019).

Within the US, UR has been an element of undergraduate education for at least a century. As described by Kinkead (2012) in her cogent historical summary, two organizations, the Research Corporation for Science Advancement (RCSA) and the National Science Foundation (NSF), provided organizing frameworks and most importantly, funding, for twentieth-century research programming. Disciplinary societies such as the American Chemical Society and Sigma Xi were certainly prominent agents promoting UR as an avenue for research development. National studies of the state of US undergraduate education (e.g. Boyer Commission 1998) also provided support for UR's expansion.

Although several colleges in the Northeast/Midwest US may claim a century or more of UR tradition, most UR historians agree that the first formally recognized UR opportunities program was created in the 1960s at the Massachusetts of Institute of Technology (MIT) (Institute Archives-MIT Libraries 2019). Since the 1960s, UR has proliferated and expanded in a variety of educational environments, and was launched as a US organized movement with the formation of the Council on Undergraduate Research (CUR) in the late 1970s.

CUR originated in 1978 with ten faculty members in chemistry-related disciplines. CUR was fueled by the zestful energy of a small group of people, a shared and inspiring vision, and driven by both challenge and opportunity. One challenge was the perception amongst many US academics that research was the province of faculty members and graduate students at research-intensive, doctoral-granting institutions. The faculty members who wished to overturn this ruling paradigm tended to be well-qualified researchers who had trained at the top research universities in the country, now pursuing faculty careers at small, predominantly undergraduate institutions (PUIs). Mostly hailing from the sciences, these faculty members pursued diverse strategies to remain active in their research programs. Early CUR leaders seized upon the opportunity to enlarge the research aims of faculty at small PUIs through expanding research activities to undergraduate students, resulting in better educational outcomes for students, and research support for faculty scholarly endeavors. These two imperatives—supporting faculty research as well as student success—arguably remain the most potent drivers of the US undergraduate research movement.

Apart from the availability of funding, which is certainly an essential factor in UR's development, I believe that three other major trends dominate UR's evolution in the U.S. First, undergraduate research's appeal has broadened to engage faculty not only within science, technology, engineering, and mathematics (STEM) fields, but also those in social and health sciences, business, and arts and humanities disciplines (Crawford et al. 2014). Increasing disciplinary diversity correlates with more academic institution-wide support for undergraduate research infrastructure, faculty and student recognition and reward systems, faculty professional development, and creation of centralized undergraduate research offices.

Second, faculty demographic changes within the last two decades—changes that will accelerate with waves of retirements of faculty who are part of the so-called baby boomer generation—mean that faculty now being hired tend to have enjoyed their own undergraduate research experiences, are from groups historically underrepresented in higher education, and wish to replicate and expand these experiences for their own students (Webber et al. 2013).

Third, research conducted by scholars such as George Kuh and colleagues at Indiana University (Kuh et al. 2007; Kuh 2008; Kuh and O'Donnell 2013); David Lopatto at Grinnell College (Lopatto 2010); Sandra Gregerman and colleagues at University of Michigan (e.g. Nagda et al. 1998; Hathaway et al. 2002); Silvia Hurtado and colleagues at University of California at Los Angeles (e.g. Hurtado et al. 2014; Eagan et al. 2013); and several other groups (Finley and McNair 2013; Laursen et al. 2010; Schmitz and Havholm 2015) have demonstrated a correlation between undergraduate research participation and student success, examining such parameters as grades, graduation rates, and transitions to graduate school or workplace. This participation is particularly impactful for students from groups historically underrepresented in higher education.

Student success—principally (but not exclusively) defined as degree completion—is currently of paramount importance in the US. Given the high cost of most undergraduate degree programs, skyrocketing student loan debt, the promise that UR offers to help close achievement gaps between majority and historically underrepresented groups, and the career advancement offered by undergraduate degree acquisition, institutional investments in undergraduate research have increased appreciably within the last decade, and are mirrored by the rapid growth of CUR's membership, programs, and services. Three brief and highly readable syntheses of literature on UR's positive impact are Crowe and Brakke (2008), Osborn and Karukstis (2009), and Altman, Chiang, Hamann, Peterson, and Orel (2019).

Given these trends, it is not surprising that present-day CUR spans all academic disciplines through its 13 divisions, engaging more than 10,000 individual members at close to 1000 academic institutions, including a growing number outside of the US. It has also enlarged its institutional membership base from the core PUI constituency to embrace most types of institutions, including doctoral-granting research intensives and community colleges. I would note that in the US, institutions are sorted into what is termed the Carnegie classification system, which is oriented primarily to levels of research activity, the numbers of students enrolled as undergraduates and/or graduates, and whether an institution is public or private (Carnegie Classification n.d.). In 2011, CUR formally merged with the National Conference on Undergraduate Research (NCUR), a more student-focused organization that had developed the infrastructure to support a student research conference attracting several thousand students yearly.

Throughout its 40-year history, the CUR's mission has been repeatedly affirmed: "to support and promote high-quality undergraduate student-faculty collaborative research and scholarship." While it would be inaccurate to state that there is a unified definition for high-quality undergraduate research shared by all institutions of higher education in the US, there is a definition formally supported by the CUR. According to the CUR (Council on Undergraduate Research n.d.-a), undergraduate research is "an inquiry or investigation conducted by an undergraduate student that makes an original intellectual or creative contribution to the discipline." Although the primacy of the concept of "originality" is strongly held within CUR, CUR members embrace a variety of modalities through which original research, scholarship, and/or creative inquiry can be accomplished. Increasingly, the strategy of curricular "scaffolding" of experiences and skill acquisition to help prepare students to achieve original research contributions has received significant focus (e.g. Nadelson et al. 2010; Shanahan 2012; Chamely-Wiik et al. 2014). The most common metric for defining originality of contribution is publication of student research, or presentations at disciplinary conferences or NCUR.

CUR's Role in the International Undergraduate Research Movement: Collaborator, Connector, and Innovator

Throughout CUR's evolution, engagement with non-US individuals and organizations has been of great importance, often effected by existing disciplinary faculty-to-faculty collaborations. In 2012, CUR's Council, its main governance body, formally enshrined internationalization of undergraduate research as one of five (5) central pillars for the organization. Three dominant engagement strategies have assisted CUR in its efforts to support and engage non-US partners. First, in addition to regularly publishing diverse authors, including those from outside US academic institutions, CUR's peer-reviewed journal, *Scholarship and Practice of Undergraduate Research (SPUR)* (formerly *CUR Quarterly*) inaugurated an International desk more than a decade ago. This special section of the journal shares international UR perspectives by non-US authors on a quarterly basis (e.g. Naipi and Airini 2019), thus expanding the cognizance of international advances in undergraduate research theory and practice among the predominantly US readership of the journal.

Second, CUR has engaged with groups such as the International Society for the Scholarship of Teaching and Learning (ISSOTL), the British Conference of Undergraduate Research (BCUR), the Australasian Council for Undergraduate Research (ACUR), and emerging coalitions of universities in Canada, Germany, and other countries, to help develop global undergraduate research infrastructure and organizational expertise. Mutually supportive activities have included joint sponsorship of professional development sessions at meetings, speaker exchanges, and promotion and marketing of non-US organizations' events to CUR membership.

Third, CUR partnered with Qatar University, BCUR, and ACUR to create the World Congress on Undergraduate Research (WorldCUR), a ground-breaking and inspiring way to create direct research connections between undergraduate scholars in different countries, centered on global challenges such as climate change, human migrations, etc. CUR's main roles have been to provide part of the abstract review, event management infrastructure, promotion, and marketing of the event. The first WorldCUR was held at Qatar University in 2016 and attracted close to 200 attendees. A retrospective look at the first WorldCUR was produced by Julio Rivera and colleagues (Rivera et al. 2018), and highlighted the empowerment and cultural capital built by attendees, as well as the opportunities for meaningful scholarly interchanges and forging of new international networks. The congress was held for a second time at the University of Oldenburg in spring 2019 (Second World Congress on Undergraduate Research 2019). Double in size in comparison to the 2016 event, and including representation from more than 30 nations, the Second WorldCUR offered diverse and innovative opportunities for emerging researchers to present and discuss their work with colleagues. Representatives from partnering organizations (CUR, BCUR, ACUR, Qatar University, and German undergraduate research programs) are now undertaking a formal bid process to identify the host of WorldCUR 2022. They have also created a formal organizing body—the Alliance for Global Undergraduate Research—which will identify roles and responsibilities associated with the triennial event, and other jointly supported international undergraduate research programs.

What Are Core Elements of a "Typical" UR Program at a US Institution?

There are over 4000 institutions of higher education in the US. About 35% of US higher education institutions are community colleges, which have less well-developed UR traditions, although that is changing (Hensel and Cejda 2014). Approximately 20% are post-secondary trade schools or specialty (e.g. Christian bible schools) institutions, which often do not offer undergraduate research.

For the remaining 2000 institutions, determining the numbers of students engaged in UR is a challenge (e.g. Wilson 2012). One entity that provides annual assessment of student engagement across a wide range of higher-education institutions is the Center for Postsecondary Study at Indiana University (Bloomington), which designs, implements, and reports on the results of an annual National Survey of Student Engagement (NSSE). The NSSE offers a comprehensive assessment of student engagement with respect to a variety of parameters, including questions on student participation in research with a faculty mentor. In 2018, the NSSE was administered at 511 institutions (489 in the US, 16 in Canada, and 6 in other countries), with 23% of student respondents reporting that they undertook research with faculty mentors by their final year of baccalaureate study. Students in the biological sciences report close to 50% participation in undergraduate research, not surprising when one considers the numbers of students applying to medical or dental school. Health professions schools, which are predominantly post-baccalaureate in the US, often regard UR participation, particularly if it leads to a publication, as one criterion for program entry (Association of American Medical Colleges 2015).

Although there exists a great diversity of approaches with respect to UR programming across the US, with significant differences between the different Carnegie classifications of institutions, there are some attributes of UR programming that are common to most US colleges and universities that engage in UR. These include, but are not limited to: some form of financial support for students and faculty who engage in UR, an administrative office of UR, and some form of recognition of UR accomplishments. In 2012, CUR published a set of guidelines (Hensel 2012) for institutions to consider when assessing their UR programs and services. Entitled *Characteristics of Excellence in Undergraduate Research*, or *COEUR*, this publication has come to serve as a basis for UR institutional assessment for many US colleges and universities.

I would submit that there are four (4) basic models for undergraduate research at US institutions: (1) the apprentice model, otherwise known as the one-on-one faculty-mentored undergraduate research experience; (2) research embedded in the curriculum; (3) community-based research, and (4) partnerships with businesses, research laboratories, or agencies. Appreciable overlap exists between these models.

The first model (apprenticeship) is arguably the most common and formally instituted at a number of institutions, particularly smaller PUIs. The individual faculty-mentored UR experience usually involves a small (less than ten) group of students, and often takes place as a summer intensive experience, between 8 and 10 weeks in duration. Many of the students who engage in the apprentice model are in at least the third year of baccalaureate study, and funding may be provided either by the institution, or by federal agency grant programs.

The second model (research embedded in the curriculum, also known as course-based undergraduate research) is now growing rapidly (e.g. Karukstis and Elgren 2007; Dolan 2016), and has the potential to engage much larger numbers of students than the traditional apprenticeship model. Of course, appreciable overlap exists between these models, with one of the fastest growing UR dimensions the first-year research experience, which is often course-based, and serves as both an opportunity for early stage research and acculturation to research methodology in specific academic disciplines (e.g. Bowman and Holmes 2018).

Community-based participatory research, which infuses elements of service learning into undergraduate research practices, is an important model for undergraduate research support (e.g. Cooke and Thorme 2011). Since 2002, the American Association of State Colleges and Universities (AASCU), an organization of approximately 400 regional comprehensive US universities (often PUIs, with graduate degrees granted at the masters' but commonly not at doctoral levels), has promoted the "stewards of place" paradigm, stating that AASCU institutions have a special mission to support regions with educational programs and services. Expanding and enhancing community-based undergraduate research is a cornerstone of AASCU's "Stewards of Place" initiative (Ambos 2015).

Partnerships with businesses and industries have resulted in a wide range of UR collaborations, many in the fields of biotechnology, cyberinfrastructure, mathematics, engineering, and health sciences. For example, Worcester Polytechnic University (WPI) in Massachusetts collaborates with corporate partners to create group UR projects as well as capstone research (Worcester Polytechnic University 2019). In their new publication on undergraduate research in mathematics, Dorff, Henrich, and Pudwell (2019) identify many industries for which partnerships with mathematics research are possible. Campuses may even serve as platforms for UR contributions, as described by Parnell, Berutich, Henn, and Koressel (2014).

A student enrolled at a US academic institution with a robust UR program would engage in a fairly predictable yearly pattern of activities. Fall semesters will often be devoted to matching faculty research mentors with students seeking UR experiences, as well as working with students who are applying to graduate programs, medical, or other health professional schools. The winter holiday hiatus often includes three weeks to month-long immersive research experiences. The spring presents continuation of mentored research experiences, with formal exhibitions of research often made in oral and/or poster form at an institution-wide research celebration day. The summer months often include 8–10 week-long intensive research experiences. Throughout the year, UR students may receive professional development such as training in research ethics, human subjects' protocols, oral and written communications, and preparing for post-baccalaureate study and/or fellowship applications.

Three Pillars of US UR Support: Funding, Policy Frameworks, and Recognition

There are three main pillars supporting US UR: (1) funding, (2) policies, and, (3) recognition and reward programs. These topics are addressed in turn through the lenses of institutional and national/regional efforts in each of these three categories.

Funding UR: Institutional Investments Leveraging Diverse External Sources

UR in the US has always been funded through diverse sources, cobbled together with precarious sustainability. Individual colleges and universities, particularly those with substantial endowments, often allocate funds from institutional resources to support UR. Most funds go toward research infrastructure (laboratories, equipment), staffing UR offices, and stipends to faculty and students engaging in research. At some institutions (e.g. University of Wisconsin Eau Claire 2019), students have agreed to have a portion of their fees dedicated to UR. At other institutions, funds raised through alumni donations are spurring UR's growth (e.g. Xia 2017).

Within the US, there are several government agencies that provide appreciable funding for undergraduate research, among them the National Science Foundation (NSF), particularly the Research Experiences for Undergraduates (REU) program; National Institutes of Health (NIH), Department of Defense, Department of Energy, Department of Education, the National Aeronautics and Space Administration, the National Oceanic and Atmospheric Administration, and several others. Of specific importance to faculty in the arts and humanities are the investments in undergraduate research supported by the National Endowment for the Humanities and the National Endowment for the Arts. In recent years, partly through CUR's advocacy, programs that identify undergraduate research as an accepted target for grant funding have become more common for NEH investments, particularly in association with the digital humanities program.

In addition, several aspects of the US financial aid infrastructure provide key assists to undergraduate research support for low-income students. Two examples are Pell grants, and federal work study (FWS). The former program, which provides grants (rather than loans), recently transitioned to a year-round, rather than a semester/quarter-based system, thus allowing low-income students to obtain Pell grant aid for course-based intensive summer research programs, the more traditional and still widely used undergraduate research program option in the US. The FWS program provides on-campus paid employment, and is becoming more commonly used to provide undergraduate research assistantships for low-income students (e.g. Nazaire and Usher 2015).

Private philanthropic foundations, such as the Andrew Mellon Foundation and Howard Hughes Medical Institute (HHMI) are also important players in the US undergraduate research movement. The Mellon foundation is currently providing substantial grant support for undergraduate research and professional advancement in the arts and humanities, including the Mellon Mays Undergraduate Fellowship program (n.d.). Among its extensive portfolio of biomedical grant programs, the HHMI developed an innovative initiative to support early undergraduate research in life science laboratories (Howard Hughes Medical Institute 2015).

Comparatively little UR support currently affiliates with states or even regions of the US. Notable exceptions include the undergraduate research programs supported by the R. J. Murdock Trust (Murdock Trust n.d.), which funds colleges and universities in the northwestern US, and the federal *Established Program to Stimulate Competitive Research* (EPSCoR—NSF) and *Institutional Development Award Networks of Biomedical Research Excellence* (IDeA INBRE—NIH) programs (Established Program to Stimulate Competitive Research n.d.; Institutional Development Award Networks of Biomedical Research Excellence n.d.), which support states that receive disproportionately low amounts of federal research support from grant programs competed nationally.

Policy Frameworks for UR: Diverse and Connected to Economic Considerations

As is the case for UR funding, policies that affect UR are highly diverse and unsystematically applied across colleges and universities. By far the best predictor for an institution's UR health is its policy framework with respect to the work that faculty do and the accessibility of UR opportunities for diverse students. Does the institution compensate faculty for mentorship of UR students? Do all students (not just those of financial means) have access to UR experiences?

If we pose the specific question: "have faculty workloads been adjusted to accommodate faculty time to supervise undergraduate research," the answer is *yes* for many STEM faculty members supported by federal grants, and at some institutions regardless of disciplines. Unfortunately, the answer is *no, or not yet* if we look at support levels at individual academic institutions, particularly for non-STEM faculty. At some institutions, sophisticated analyses of faculty assignments, and how they may be adjusted to accommodate faculty UR mentorship have been undertaken (e.g. Barthell et al. 2013; Free et al. 2015; Morrison et al. 2018). Several essays in Hensel and Paul's 2012 publication, *Faculty Support and Undergraduate Research: Innovations in Faculty Role Definition, Workload, and Reward* highlight specific strategies (e.g. Paul 2012) that institutions have adopted to account for the time that faculty mentors spend with research students. Common strategies are: "banking" independent study units to allow for reduction in teaching assignments in a future year, stipend payments to faculty who are mentoring undergraduate researchers, particularly in summer months, and increasing the individual course credit (often called weighted teaching units) toward faculty workload assignments.

The roles of state and/or federal governments to create UR-friendly policies are ambiguous at best, with some notable (and positive) exceptions. The aforementioned decision for Pell grants to be available year-round for low-income students promises to help close UR participation gaps relating to socioeconomic status. The foremost role of federal agencies has been to respond to the directions set by national panels convened by trusted, high-status organizations such as the National Academies of Sciences, Engineering, and Medicine (e.g. National Academies of Sciences, Engineering, and Medicine 2015, 2017; National Research Council 2003, 2012). For example, the most recent comprehensive study of UR benefits was undertaken by a commission of the National Academies of Sciences, Engineering, and Medicine from 2015 to 2017. Culminating in a publicly available report entitled *Undergraduate Research Experiences for STEM Students: Successes, Challenges, and Opportunities*, the commission both affirmed the high value of undergraduate research for student success, and called for more rigorous, quasi-experimental studies of UR's impact. The effect of this report on federal grant funding expectations is to raise the level of sophistication of studies of UR's positive impact on student achievement.

Several documents issued by US-government-sponsored studies encourage innovation to support economic development and job creation, and directly connect economic development with expanded UR support. One recent US government publication that has been of particular significance to national movements to expand UR practices is the February 2012 report from the President's Council of Advisors on Science and Technology (PCAST): *Engage to Excel: Producing One Million Additional College Graduates with Degrees in Science, Technology, Engineering, and Mathematics*. In this report, PCAST affirmed the imperative to grow the nation's STEM workforce, particularly through support of students from groups historically underrepresented in STEM, and identifies UR as a key strategy to achieve this goal.

Recognition and Rewards for UR: Faculty, Students, and Institutions

Recognition of UR achievement is one of the fastest growing dimensions of the US UR movement. Faculty and student recognition avenues are viewed primarily through the lens of individual professional advancement. For faculty members, advancement is commonly connected to retention and promotion, including tenure evaluation. For students, advancement is primarily connected to preparation for the workforce and/or post-baccalaureate study. Institutions tend to value recognitions that highlight their national stature and degree value.

Faculty Recognition and Rewards Faculty members' contributions as UR mentors exist at many US institutions, but there is quite a bit of variance, and many faculty members would argue that recognition is not always valued proportional to faculty effort (e.g. Morrison et al. 2018). One of the most important ways that faculty can be recognized is through identification of UR mentorship as a component of faculty professional achievement, including such activities as revising curriculum to emphasize research and promoting student coauthors on publications. The concept of *tenure*, whereby an individual faculty member can achieve an academic appointment of indefinite duration, is a key milestone for US faculty for whom this is available. Institutions that have modified tenure and/or promotion requirements to identify UR mentorship (e.g. Barthell et al. 2013) have had appreciable success in motivating faculty to engage in mentorship. Many institutions also formally acknowledge faculty UR mentors at yearly award programs (e.g. University of Washington n.d.).

Over the past decade, CUR has instituted a number of faculty mentorship awards (Council on Undergraduate Research n.d.-b) that carry both monetary and symbolic value, and partnered with prominent organizations such as the Goldwater Scholarship Foundation to reward and celebrate faculty UR mentorship.

Student Recognition and Rewards Particularly in the sciences, student coauthorship and copresentation with faculty are frequent elements of undergraduate research practice. In my opinion, three of the most important drivers for increasing incidence of student coauthored papers are professional and graduate school expectations, fellowship requirements (e.g. Fulbright Scholars), and changes in faculty retention and promotion policies at some campuses to provide recognition of student coauthors. Many colleges and universities have in-house journals for publishing students' UR projects. CUR's *Scholarship and Practice of Undergraduate Research* offers the *Undergraduate Research Highlights* section, in which synopses of faculty–student coauthored papers from peer-reviewed disciplinary journals are showcased.

National conferences such as NCUR, and regional and/or state-based recognition events (e.g. Wohlers et al. 2012; Swift et al. 2012; Freund and Schneider 2019) offer important avenues for presentation that are specifically tailored to students' professional development needs. CUR also holds a fall symposium, which highlights the achievements of students who have participated in NSF REU programs.

Institutional Recognition and Rewards Many institutions within the US have undergraduate research celebration days, and through the Council on Undergraduate Research's initiative, the concept of a national undergraduate research week has taken hold, and now attracts many examples of undergraduate research celebrations each spring (e.g. University of Utah 2019). CUR's Posters on the Hill celebration, now in its 23rd year, showcases top undergraduate researchers at a poster event on Capitol Hill, attracting an audience of legislators and federal funders. Many states and regions have research days during which state legislators interact with student researchers and faculty mentors, for example, Wisconsin's "Posters in the Rotunda" (Knutsen 2016). Last, but not least, CUR created an institutional award program based on the CUR guidelines. Termed AURA for Campus-wide Award for Undergraduate Research Achievement (Council on Undergraduate Research n.d.-b), the program has attracted many applications from institutions, and made 11 awards in the last four years.

What Are the Major Challenges and Opportunities Ahead for US Undergraduate Research?

Although the stock answers to this questions are always "time and money," CUR's experience, particularly as illumined through a series of grants funded through the US National Science Foundation, is that disciplinary, departmental, and institutional cultures ultimately play crucial roles in determining whether undergraduate research will expand, or stagnate, on a particular campus (Malachowski et al. 2015). In particular, professional advancement and policy frameworks for faculty and students must be intentionally reengineered to reward UR participation, or UR practices cannot be sustained.

Persistent challenges with respect to stable federal research funding, undergraduate education affordability issues, rapid transitions in faculty and student demographics, and slow implementation of effective policies to support undergraduate research mentorship certainly exist. In the face of these challenges, however, I believe that US UR can and must expand vigorously in innovative and far-reaching ways. In the future, we will likely see more detailed and quantitative studies of UR's impact, better-defined incentives for low-income students to participate in undergraduate research, and UR programmatic growth at diverse institutions, particularly community colleges. Other future developments that will aid in UR's expansion include better mentorship training for current and future (i.e. graduate students) faculty members. Closer connections between undergraduate research programs and community, business, and industry needs, and increasing international collaborations will provide a more stable base of fiscal support and community appreciation. In the twenty-first century, undergraduate education will be dominated by the concept of the value of the credential obtained—to the individual, to employers, and to society. UR has a unique power to rapidly expand a nation's research and innovation capacity, and solve pressing societal and environmental challenges that affect all.

References

Altman, J., Chiang, T.-M., Hamann, C. S., Makhluf, H., Peterson, V., & Orel, S. (2019). *Undergraduate research: A road map for meeting future national needs and competing in a world of change*. Council on undergraduate research white paper #1, Washington, DC. Retrieved from https://www.cur.org

Ambos, E. (2015). Promoting student engagement. In R. Dunfee & A. Vaidya (Eds.), *Operationalizing stewards of place: Implementing regional engagement and economic development strategies* (pp. 13–31). Washington, DC: American Association of State Colleges and Universities. Retrieved from https://aascu.org/freepubs/OperationalizingStewardsofPlace.pdf

Association of American Medical Colleges. (2015). *Research experience for prospective Ph.D. candidates.* Retrieved from https://students-residents.aamc.org/applying-medical-school/article/research-experience-prospective-phd-candidates/

Barthell, J., Chen, W. R., Endicott, B. K., Hughes, C. A., Radke, W. J., Simmons, C. K., & Wilson, G. M. (2013). Encouraging and sustaining a culture of student-centered research at a predominantly undergraduate institution. *CUR Quarterly, 34*(1), 41–47.

Bowman, N. A., & Holmes, J. M. (2018). Getting off to a good start? First-year undergraduate research experiences and student outcomes. *Higher Education, 76*(1), 17–33. https://doi.org/10.1007/s10734-017-0191-4.

Boyer Commission on Educating Undergraduates in the Research University. (1998). *Reinventing undergraduate education: A blueprint for America's research universities.* Stony Brook.

Carnegie Classification of Institutions of Higher Education. (n.d.). Retrieved from http://carnegieclassifications.iu.edu/

Chamely-Wiik, D., Dunn, K., Heydet-Kirsch, P., Holman, M., Meerof, D., & Peluso, J. (2014). Scaffolding the development of students' research skills for capstone experiences: A multi-disciplinary approach. *CUR Quarterly, 34*(3), 18–25.

Cooke, D., & Thorme, T. (2011). *A practical handbook for supporting community-based research with undergraduate students.* Washington, DC: Council on Undergraduate Research.

Council on Undergraduate Research. (n.d.-a). *Mission.* Retrieved from https://www.cur.org/who/organization/mission/

Council on Undergraduate Research. (n.d.-b). *Awards.* Retrieved from https://www.cur.org/what/awards/

Crawford, I., Orel, S. E., & Shanahan, J. O. (Eds.). (2014). *How to get started in arts and humanities research with undergraduates.* Washington, DC: Council on Undergraduate Research.

Crowe, M., & Brakke, D. (2008). Assessing the impact of undergraduate research experiences on students: An overview of current literature. *CUR Quarterly, 28*(1), 43–50.

Dolan, E. L. (2016). *Course-based undergraduate research experiences: Current knowledge and future directions.* Paper commissioned for the committee on strengthening research experiences for undergraduate STEM students, Board on Science Education, Division of Behavioral and Social Sciences and Education,

Board on Life Sciences, Division of Earth and Life Studies. Retrieved from https://sites.nationalacademies.org/cs/groups/dbassesite/documents/webpage/dbasse_177288.pdf

Dorff, M., Henrich, A., & Pudwell, L. (2019). *A mathematician's practical guide to mentoring undergraduate research.* Washington, DC: Published Jointly by the Mathematical Association of America, the American Mathematical Society, and the Council on Undergraduate Research.

Eagan, M. K., Jr., Hurtado, S., Chang, M., Garcia, G. A., Herrera, F. A., & Garibay, J. C. (2013). Making a difference in science education: The impact of undergraduate research programs. *American Educational Research Journal, 50*(4), 463–713.

Established Program to Stimulate Competitive Research. (n.d.). Retrieved from https://www.nsf.gov/od/oia/programs/epscor/

Finley, A., & McNair, T. (2013). *Assessing underserved students' engagement in high-impact practices.* Washington, DC: Association of American Colleges and Universities.

Free, R., Griffith, S., & Spellman, B. (2015). Faculty workload issues connected to undergraduate research. In M. Malachowski, J. M. Osborn, K. K. Karukstis, & E. L. Ambos (Eds.), *Enhancing and expanding undergraduate research: A systems approach* (Vol. 169, pp. 51–60). San Francisco: New Directions for Higher Education, Jossey-Bass.

Freund, E. C., & Schneider, K. R. (2019). Establishing a statewide celebration of undergraduate research: History and lessons learned. *Scholarship and Practice of Undergraduate Research, 2*(3), 28–34. https://doi.org/10.18833/spur/2/3/3.

Hathaway, R. S., Nagda, B., & Gregerman, S. (2002). The relationship of undergraduate research participation to graduate and professional education pursuit: An empirical study. *Journal of College Student Development, 43*(5), 614–631.

Hensel, N. H. (Ed.). (2012). *Characteristics of excellence in undergraduate research.* Washington, DC: Council on Undergraduate Research. Retrieved from https://www.cur.org/assets/1/23/COEUR_final.pdf

Hensel, N. H., & Cejda, B. (Eds.). (2014). *Tapping the potential of all: Undergraduate research at community colleges.* Washington, DC: Council on Undergraduate Research.

Hensel, N. H., & Paul, E. L. (Eds.). (2012). *Faculty support and undergraduate research: Innovations in faculty role definition, workload, and reward.* Washington, DC: Council on Undergraduate Research.

Howard Hughes Medical Institute. (2015, April 28). *Students break new ground in understanding genetic diversity of bacteriophages.* Retrieved from https://www.hhmi.org/news/students-break-new-ground-understanding-genetic-diversity-bacteriophages

Hurtado, S., Eagen, K., Figueroa, T., & Hughes, B. (2014). *Reversing underrepresentation: The impact of undergraduate research programs on enrollment in STEM graduate programs*. Retrieved from https://www.heri.ucla.edu/nih/downloads/AERA-2014-Undergraduate-Research-And-STEM-Grad-Enrollment.pdf

Institute Archives, Massachusetts Institute of Technology Libraries. (2019). Retrieved from https://libraries.mit.edu/mithistory/institute/offices/undergraduate-research-opportunities-program-urop/

Institutional Development Award Networks of Biomedical Research Excellence. (n.d.). Retrieved from https://www.nigms.nih.gov/Research/DRCB/IDeA/pages/INBRE.aspx

Karukstis, K. K., & Elgren, T. E. (Eds.). (2007). *Designing and sustaining a research-supportive curriculum: A compendium of successful practices*. Washington, DC: Council on Undergraduate Research.

Kinkead, J. (2012). What's in a name? A brief history of undergraduate research. *CUR Quarterly, 33*(1), 20–29.

Knutsen, K. (2016, April 14). *UW day at the capitol, posters in the Rotunda celebrated*. Retrieved from https://news.wisc.edu/uw-day-at-the-capitol-posters-in-the-rotunda-celebrated/

Kuh, G. D. (2008). *High-impact educational practices: What they are, who has access to them, and why they matter?* Washington, DC: Association of American Colleges and Universities.

Kuh, G. D., & O'Donnell, K. (2013). *Ensuring quality & taking high-impact practices to scale*. Washington, DC: Association of American Colleges and Universities.

Kuh, G. D., Kinzie, J., Cruce, T., Shoup, R., & Gonyea, R. M. (2007). *Connecting the dots: Multi-faceted analyses of the relationships between student engagement results from the NSSE, and the institutional practices and conditions that foster student success*. Bloomington: Indiana University Center for Postsecondary Research. Retrieved from http://nsse.iub.edu/pdf/Connecting_the_Dots_Report.pdf

Laursen, S., Hunter, A.-B., Seymour, E., Thiry, H., & Melton, G. (2010). *Undergraduate research in the sciences: Engaging students in real science*. San Francisco: Jossey-Bass.

Lopatto, D. (2010). *Science in solution: The impact of undergraduate research on student learning*. Washington, DC/Tucson: Council on Undergraduate Research/Research Corporation for Science Advancement.

Malachowski, M., Osborn, J. M., Karukstis, K. K., Ambos, E. L., Kincaid, S. L., & Weiler, D. (2015). Fostering undergraduate research change at the system and consortium level: Perspectives from the Council on Undergraduate Research. In M. Malachowski, J. M. Osborn, K. K. Karukstis, & E. L. Ambos (Eds.), *Enhancing and expanding undergraduate research: A systems approach* (Vol.

169, pp. 95–106). San Francisco: New Directions for Higher Education, Jossey-Bass.

Mellon Mays Undergraduate Fellowship Program. (n.d.). Retrieved from https://www.mmuf.org/

Mieg, H. A. (Ed.). (2019). *Inquiry-based learning – Undergraduate research: The German multidisciplinary experience*. Cham: Springer. https://doi.org/10.1007/978-3-030-14223-0.

Morrison, J. A., Berner, N. J., Manske, J. M., Jones, R. M., Davis, S. N., & Garner, P. W. (2018). Surveying faculty perspectives on undergraduate research, scholarship, and creative activity: A three-institution study. *Scholarship and Practice of Undergraduate Research, 2*(1), 43–54. https://doi.org/10.18833/spur/2/1/1.

Murdock Trust. (n.d.). *Undergraduate colleges and universities*. Retrieved from https://murdocktrust.org/grant-opportunities/undergraduate-colleges-universities/

Nadelson, L., Walters, L., & Waterman, J. (2010). Course integrated undergraduate research experiences structured at different levels of inquiry. *Journal of STEM Education: Innovations and Research, 11*, 27–44.

Nagda, B. A., Gregerman, S. R., Jonides, J., von Hippel, W., & Lerner, J. S. (1998). Undergraduate student faculty research partnerships affect student retention. *Review of Higher Education, 22*, 55–72.

Naipi, S., & Airini. (2019). Knowledge makers: Indigenous undergraduate researchers and research. *Scholarship and Practice of Undergraduate Research, 2*(3), 52–60. https://doi.org/10.18833/spur/2/3/7.

National Academies of Sciences, Engineering, and Medicine. (2015). *Integrating discovery-based research into the undergraduate curriculum: Report of a convocation*. Washington, DC: National Academies Press.

National Academies of Sciences, Engineering, and Medicine. (2017). *Undergraduate research experiences for STEM students: Successes, challenges, and opportunities*. Washington, DC: The National Academies Press. https://doi.org/10.17226/24622.

National Research Council. (2003). *Evaluating and improving undergraduate teaching in science, technology, engineering, and mathematics*. Washington, DC: National Academy Press.

National Research Council. (2012). *Discipline-based education research: Understanding and improving learning in undergraduate science and engineering*. Washington, DC: The National Academies Press.

National Survey of Student Engagement. (2018). *Engagement insights: Survey findings on the quality of undergraduate education – Annual results 2018*. Bloomington: Indiana University Center for Postsecondary Research. Retrieved from http://nsse.indiana.edu/html/annual_results.cfm

Nazaire, D. W., & Usher, B. M. (2015). Leveraging Federal Work-Study to support undergraduate research. *CUR Quarterly, 36*(2), 9–17.

Osborn, J. M., & Karukstis, K. K. (2009). The benefits of undergraduate research, scholarship, and creative activity. In M. Boyd & J. Wesemann (Eds.), *Broadening participation in undergraduate research: Fostering excellence and enhancing the impact* (pp. 41–53). Washington, DC: Council on Undergraduate Research.

Parnell, R., Berutich, L., Henn, A., & Koressel, N. (2014). The campus as a four-year undergraduate learning laboratory on sustainability: Linking facilities, operations, curriculum and community engagement. *CUR Quarterly, 35*(1), 11–19.

Paul, E. L. (2012). New directions for faculty workload models: Focusing on high-impact learning practices. In N. Hensel & E. L. Paul (Eds.), *Faculty support and undergraduate research: Innovations in faculty role definition, workload, and reward* (pp. 133–145). Washington, DC: Council on Undergraduate Research.

President's Council of Advisors on Science and Technology (PCAST). (2012). *Engage to excel: Producing one million additional college graduates with degrees in science, technology, engineering, and mathematics.* Washington, DC: Office of Science and Technology Policy, Executive Office of the President. Retrieved from https://obamawhitehouse.archives.gov/sites/default/files/microsites/ostp/pcast-engage-to-excel-final_2-25-12.pdf

Rivera, J., Khelifa, M., Hamdah, B. A., Al-Hamadi, A. M., & Zdgiebloski, E. S. (2018). A global conversation: Reflections from the first world congress on undergraduate research. *Scholarship and Practice of Undergraduate Research, 2*(1), 55–62. https://doi.org/10.18833/spur/2/1/4.

Schmitz, H. J., & Havholm, K. (2015). Undergraduate research and alumni: Perspectives on learning gains and post-graduation benefits. *CUR Quarterly, 35*(3), 15–22.

Second World Congress on Undergraduate Research. (2019). Retrieved from https://uol.de/en/worldcur2019/

Shanahan, J. (2012). Building undergraduate research into the curriculum. In N. Hensel & E. Paul (Eds.), *Faculty support and undergraduate research: Innovation in faculty role definition, workload, and reward* (pp. 68–76). Washington, DC: Council on Undergraduate Research.

Swift, J. N., Ambos, E. L., Swift, C., & Ash, C. (2012). A regional undergraduate research conference comes of age in southern California. *CUR Quarterly, 32*(4), 21–27.

University of Utah. (2019, April 5). *Happy undergraduate research week!* Retrieved from https://our.utah.edu/home-page/happy-undergraduate-research-week-2019/

University of Washington. (n.d.). *Research mentor awardees*. Retrieved from http://www.washington.edu/undergradresearch/symposium/mentor-awardees/

University of Wisconsin Eau Claire. (2019). *Students: Marks of excellence*. Retrieved from https://www.uwec.edu/ORSP/students/

Webber, K. L., Nelson Laird, T., & BrckaLorenz, A. (2013). Student and faculty engagement in undergraduate research: Evidence from NSSE and FSSE. *Research in Higher Education, 54*(2), 227–245.

Wilson, A. (2012). Using the National Survey of student engagement to measure undergraduate research participation. *CUR Quarterly, 32*(3), 9–14.

Wohlers, T. E., Wilson, G. M., Rao, N., Gwyn, L., Schoonover-Redden, K., & Malayer, J. R. (2012). Celebrating undergraduate research in Oklahoma: The history and impact of statewide undergraduate research conferences. *CUR Quarterly, 32*(4), 15–20.

Worcester Polytechnic University. (2019). *Research partnerships*. Retrieved from https://www.wpi.edu/research/partnerships

Wulf, A. (2015). *The invention of nature: The adventures of Alexander von Humboldt, the lost hero of science*. London: John Murray.

Xia, R. (2017). Cal Poly San Luis Obispo receives $110 million, the largest private donation in CSU history. *Los Angeles Times*. Retrieved from https://www.latimes.com/local/lanow/la-me-cal-poly-slo-donation-20170502-story.html

CHAPTER 3

Turning a Dream into Reality: Building Undergraduate Research Capacity Across Australasia

Angela Brew and Lilia Mantai

Introduction

In this chapter, we focus on Australia where undergraduate research has gained traction in the last decade. It has emerged as a significant force in higher education more recently than is the case in the United States (US). To paint the picture of the Australian landscape of undergraduate research, we introduce a few significant initiatives that have informed and promoted undergraduate students' research opportunities and its value across the country. Specifically, we present a ten-year program of educational

A. Brew (✉)
Office of Deputy Vice-Chancellor (Research), Macquarie University, Sydney, NSW, Australia
e-mail: angela.brew@mq.edu.au

L. Mantai
Academic Lead Course Enhancement, University of Sydney, Sydney, NSW, Australia
e-mail: lilia.mantai@sydney.edu.au

N. H. Hensel, P. Blessinger (eds.), *International Perspectives on Undergraduate Research*,
https://doi.org/10.1007/978-3-030-53559-9_3

development and enhancement initially designed to develop undergraduate research and research-based learning in a large Australian research-intensive university (approximately 39,000 students) and later to promote and support undergraduate research in Australasian universities more generally.

The program aimed to increase opportunities for students to engage in research within the undergraduate curriculum and in co-curricular programs, and to enhance performance and showcase best practices where curriculum is informed by research. It aimed to enhance practice in equipping students with research skills and critical thinking through exposure to research problems and authentic research environments. The program followed a conceptual/evidence-based approach to developing inclusive, scholarly and knowledge-building communities (Brew 2006). The chapter discusses the challenges we faced and highlights opportunities our experience offers to inform the development of undergraduate research engagement elsewhere. The findings are presented in the context of the relevant literature and reference is made to what can be learned from the existing scholarship in undergraduate research where appropriate.

The Australian Context

The Australian higher education system has 43 universities, of which two are privately owned and funded. There is a mix of large research-intensive institutions which tend to be located in cities, smaller regional institutions, and former institutes of technology all of which may have pockets of research-intensity depending on their location and focus. In addition, there are 135 higher education providers mainly of a specialist nature. Overall, the system enrolls approximately 1.4 million students of which 91 per cent are in universities and 29 per cent are from overseas (Australian Government 2017).

It is worth noting that the land mass of Australia is similar to the US (excluding Alaska and Hawaii). However, the population of Australia currently stands at 24 million (as contrasted to the US's 330 million). It is also important to note that the geographical position of Australasia in the Southern Hemisphere creates an academic year which runs from March to December. This factor is important when it comes to the provision of undergraduate research opportunities within the Australian summer vacation which runs from December to February.

Undergraduate Research in Australia in 2009

In 2009, an investigation of the state of undergraduate research provision was carried out by an undergraduate researcher (Jewell and Brew 2010). It identified approximately 1500–2000 students engaged in co-curricular undergraduate research scholarship programs annually that were mostly carried out during the summer vacation (i.e. December to February), for a maximum of ten weeks. One university engaged students in term time but that was unusual. Twenty-three universities had one or more undergraduate research scholarship program. Five universities had the whole of institution schemes, but otherwise, participating students were mostly from STEM and health-related subjects. Students were selected on the basis of demonstrated academic merit and motivation to do research. They were all paid a tax-free stipend which varied depending on the length of time. Thirty-one private external organisations that were funding undergraduate research scholarships were identified. No government funding or support was forthcoming. However, the report suggested that the numbers were growing. Curriculum-based undergraduate provisions such as research-based learning courses were not surveyed at that time.

This research was carried out as part of an Australian Learning and Teaching Committee (ALTC) National Teaching Fellowship designed to bring together a team of international experts and leading Australian collaborators to provide an overview of undergraduate research practice and explore implications for the future in a series of events for academics and university leaders and managers. Study tours overseas and interviews with key individuals enabled the identification of resources and information. A website: http://www.undergraduateresearchAustralia.com was established and regional roundtables were held in different Australian states. The first two-day *Australian Summit on the Integration of Research, Teaching and Learning* was held with 90 delegates, many of whom were senior representatives of 35 universities and other bodies with an interest in undergraduate research such as the Australian Universities Quality Agency (AUQA), the Australian Learning and Teaching Committee, The Australian Council of Deans of Science, and the National Union of Students, resulted in the development of a Communiqué addressed to political leaders. The Summit identified funding for undergraduate research experience programs as a major challenge for the future.

The ALTC Fellowship identified a network of people within Australia and New Zealand who were interested in undergraduate research. A newsletter (URNA) was established to keep them informed of developments. The Fellowship work was also disseminated through presentations, an edited book, journal articles, media coverage, and conference presentations during 2010/11. The report of the external evaluator concluded:

> The Fellowship has made a very significant contribution to moving the undergraduate research agenda forward at a national level in Australia and to making the benefits of strong links between research and teaching better understood. (Brew 2010, p. 41)

Towards Inclusive Higher Education

While the ALTC Fellowship marks a significant milestone in the development of undergraduate research in Australasia, it did not stand alone. It was itself possible because it was able to stand on the shoulders of numerous researchers in a number of countries exploring the relationship between teaching and research and implications for students' engagement in research. Indeed, at the time, researchers demonstrated a growing interest in developing opportunities for undergraduates to engage in research (see e.g. Healey and Jenkins 2009; Levy and Petrulis 2012; Beckman and Hensel 2009), establishing ideas about the scholarship of learning and teaching (e.g. Hutchings 2002) and developing understanding of different forms of undergraduate research engagement, its benefits and student gains (cf. Laursen et al. 2010; Lopatto 2009).

Well-established networks in the US such as the Council on Undergraduate Research (CUR), provided numerous exemplars of good practice as well as inspiration (cf. Boyer Commission 1999; Karukstis and Elgren 2007). In Australia, John Willison and colleagues (2007) developed a conceptual model called the Research Skill Development Framework, which has been an influential approach to nurture research skills of all students in many undergraduate degrees.

Initially, our focus of attention was on short-term apprenticeship-style undergraduate research opportunities carried out principally in vacations as appeared at the time to be dominant in the US. However, while a few Australian universities have summer and winter undergraduate research programs organised either centrally or in faculties and departments, in Australia, most undergraduate research projects are completed in the final 'Honours' year. Zimbardi and Myatt (2014) identified four kinds of undergraduate research engagement in one Australian university:

apprenticeship-style, industry project, inquiry project, and a methods course. They acknowledged that these may be mixed. Several universities have special research schemes for high achieving students.

As well as a mixture of different patterns of engagement, there are also varieties in the language used to describe undergraduate research. Research-based learning and inquiry-based learning are commonly used interchangeably for undergraduate research while recognising that students may engage in research in a variety of ways (see e.g. Brew and Mantai 2017). With this in mind, we found it important to extend Beckman and Hensel's (2009) definition of undergraduate research to fit the Australian context:

> An inquiry or investigation or a research-based activity conducted by an undergraduate student that makes an original intellectual or creative contribution to the discipline and/or to understanding. (Brew 2010 following Beckman and Hensel 2009, p. 40. Our additions underlined)

Through all of this work it became clear that there are many varieties and forms of undergraduate research experience and that it exists at the level of the individual academic as well as at school or faculty level and at university level (Smith and Rust 2011). This is clearly recognised in the US Council on Undergraduate Research whose characteristics of excellence in undergraduate research (Hensel 2012) encompass all aspects of university functioning. However, this challenges all levels and parts of the university, both teaching and research, institutional strategies and objectives.

There is no escaping the fact that developing undergraduate engagement in research represents a transformative agenda. Fundamental assumptions about the role of undergraduate students in the university, the nature of knowledge and who is to be involved in generating it, as well as the relationships of students to university research, all require attention. Such challenges are embedded in ideas about students as producers (Neary 2010), students as partners (Matthews 2018; Healey et al. 2016) and students as change agents (Kay et al. 2010), where hierarchies are questioned and students' relationships with academics change. Such initiatives support a more inclusive higher education; one in which students take a role as actively participating in the academic project of the university.

These ideas are embedded in a model of universities as inclusive scholarly knowledge-building communities (Brew 2006). This model grew from investigations of research-teaching integration and practices of undergraduate research engagement in different countries. It suggests a transformation in higher education which enables students as partners, as producers and/or as change agents to flourish. The hexagonal model has six interlinked facets: research, teaching and learning, knowledge, inclusivity, community and scholarship all of which require attention (see Fig. 3.1).

The dotted lines suggest that the boundaries of the facets are not distinct, but merge into each other. The lines extend beyond the hexagon to suggest that universities are interlinked into society and accountable to, for example, industry, professions, family and media (Brew 2006).

Following the initial ALTC Fellowship, our development work has been underpinned by a transformative agenda and the values and aspirations of developing inclusive scholarly knowledge-building communities. It is fair to say that we did not consciously use the model as a guide for implementation. However, we have found it instructive to utilise it in critically analysing the subsequent spread of undergraduate research in Australasia, specifically our seven-year project (2011–2017) and our

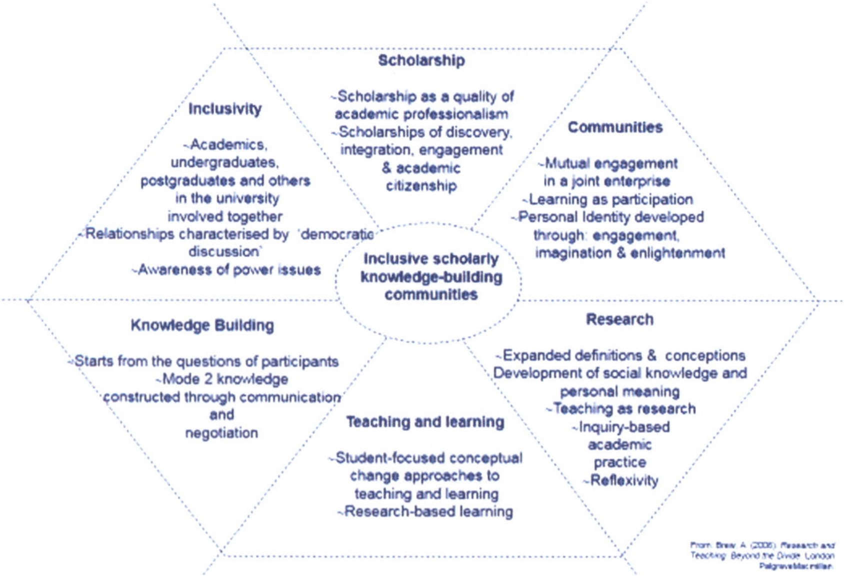

Fig. 3.1 Hexagonal model of universities as inclusive scholarly knowledge-building communities (Brew 2006, p. 32)

current work (2018–2020) in this space. As we worked in the six domains of the hexagonal model, we use these headings in this chapter to trace undergraduate research development in Australia.

2011–2017: Laying Foundations to Build Undergraduate Research Capacity in Australia

Research

A key aspect of our development program was the integration of research projects where we worked with undergraduate and postgraduate students to carry out research into existing practice. A broad view of research was taken as suggested by the above model.

Research was carried out by an undergraduate using a similar questionnaire to that used by Turner, Wuetherick and Healey (2008) and Spronken-Smith et al. (2013), to interview 200 students about their awareness of research. Interviewees were from all faculties of a large Australian metropolitan research-intensive university. Echoing previous studies, the findings suggested that these students had little awareness of research in the university (Hajdarpasic et al. 2015).

As course outcomes were available online, we were able to search the learning outcomes to see to what extent research, inquiry and related activities were a focus. In fact, despite the university's stated aim to have a research-rich environment for students, very few learning outcomes mentioned any aspects of research. This suggested that the university's broad aim was not being implemented at the course level.

We analysed photographs of noticeboards, corridors and the campus more generally and found that research was not very visible. This possibly contributed to our finding that students lacked awareness of research. We encouraged greater visibility of research, for example, corridors showcasing research posters, space made available to advertise undergraduate research opportunities, and activities etc., and this resulted in some improvements (Popenici and Brew 2013).

We interviewed 20 academics who were engaged in developing research-based learning, to identify their perceived constraints to implementation. Interestingly, we found they had varying definitions of undergraduate research which appeared to have led to different forms of student engagement ranging from atomistic and uncoordinated work, such as students engaging in a range of unrelated projects to develop specific aspects of research, to holistic and integrated practices where students are integrated into the scholarly community and engage in complete projects from inception through to publication (Brew and Mantai 2017).

The earlier survey of provision across Australian universities interviewed undergraduate research coordinators by telephone. This work identified a significant gap between student demand and the availability of research experiences across the system (Brew and Jewell 2012).

To inform policy and practice, expand staff conceptions of undergraduate research and contribute to discussions of future needs, the findings of all of these investigations were disseminated in staff development programs, university committees, conferences and journal articles.

Teaching and Learning

A staff development program was therefore designed to disseminate and discuss research findings, enhance academics' appreciation of undergraduate research potential, and contribute to curriculum change initiatives. It consisted of:

1. A university-wide working group (with staff and student representatives nominated by the 30 or so heads of department), which met over a three-year period to promote and communicate departmental activities. Issues discussed included research skills development, decision-making, ethical issues, standards, and best practice in research-based learning.
2. Central workshops and showcases, on such topics as using research in capstone courses, supervising undergraduate research, and designing research-based undergraduate courses.
3. An online course on implementing undergraduate research built using the website resources at http://www.undergraduateresearchaustralia.com offered through the university's learning management system was made available to staff.
4. Informal discussions were facilitated face-to-face and online, for example, via a wiki.

This program was evaluated through records of working group meetings reported in university committees; workshops, events and resources formally evaluated through exit and follow-up surveys; critical reflections of the working group, and scholar/ambassadors (see below); and the project leaders reported through regular meetings.

One of the key lessons for us in this program is the importance of engaging students in discussions of curriculum, in pedagogical research, and in encouraging students to demand changes in their education. As Hutchings (2002) suggests, students should be engaged in the scholarship of teaching and learning.

Knowledge Building

Knowledge building takes place within 'transaction spaces' (Nowotny et al. 2001, p. 103). This involves academics, students and others within the academy working with other professionals and interested people and groups so that new ideas can emerge. As mentioned earlier, discussions during the ALTC Fellowship, identified a need to spread knowledge of undergraduate research across the broader Australasian community. With this in mind, in 2012 we organised the inaugural Australasian Conference of Undergraduate Research (ACUR). The event was overwhelmingly successful with 130 presentations given by undergraduates on topics ranging across all disciplines; a pattern that was repeated when we organised the second conference the following year. These conferences attracted numerous student volunteers, sponsorship, academic reviewers, and comments on social media. The best papers were subsequently published in an undergraduate research journal (https://studentjournal.mq.edu.au/).

Following these two conferences, a grant from the Australian Government's Office for Learning and Teaching was obtained to make the conferences sustainable in the longer term. A Steering Group consisting of representatives from Australian and New Zealand universities was established, linked to a community of individuals developing undergraduate research worldwide. This provided a base for disseminating information and a forum for establishing documentation ensuring the quality of future conferences. Through this process the Australasian Council for Undergraduate Research (ACUR) was born (see below).

Since that time, Australasian conferences of undergraduate research have been held annually in different Australian universities. Steering Group members work within their institutions in a variety of ways to develop undergraduate research communities and to prepare students for conference attendance.

Sector-wide appreciation of the value of undergraduate research began to be developed through the 2009 Summit. We endeavoured to build on this by organising a Posters in Parliament event in 2014 in Parliament House Canberra (based on the UK and the US experience), to showcase high-quality undergraduate research to key figures and representatives from research funding bodies as well as members of the House of Representatives and Senators and to spread knowledge of undergraduate research across the broader Australian community. We saw this as a way of educating policymakers about the quality of research that undergraduates can achieve and once again highlighting the need for funding.

Judging by written comments made by attendees, and the large number of emails of support received from Vice-Chancellors, other senior university officials and politicians, this event certainly had national impact in raising the profile of undergraduate research. It also had significant impact on the student participants themselves. Students told us of their newly gained confidence and motivation and their follow-up email correspondence with various key figures as a result of the event. One undergraduate presenter who went on to do a PhD wrote:

> Both [federal politicians] showed great interest in my project, as both had a personal connection with [topic] and were aware of how greatly it affects the wider community. … [One] was incredibly encouraging, and adamant that I send him a copy of my published paper as soon as I can. In all, I hope the event will continue in the future, as it will help draw further connections, interactions, and initiatives between students, universities, politicians, and the government.

Owing to the high cost of the event and the lack of further funding, to date we have been unable to arrange further Posters in Parliament events. Other ways to promote and support undergraduate research are needed. We recognised that this was just the start of a process and that much more needed to be done.

Inclusivity

Many university educational development initiatives focus on aspects of inclusivity, for example, internationalisation, widening participation, indigenisation and entrepreneurship. It is important that such initiatives not only encourage inclusivity to be developed but that they also embody inclusive practices. In our project we tried to take an inclusive approach to development in stimulating scholarly activity amongst students, treating those researching with us as participating colleagues, and encouraging them to become ambassadors for change. This involved students as full participants at all times: investigating practice, implementing the staff development program, and in presenting research in their universities and in parliament.

Our program included undergraduate pedagogical research internships. By encouraging undergraduates to devise projects of relevance to their particular departments, we intended to grow a community of undergraduate scholar/ambassadors. Projects were supervised by departmental academics, and students worked in partnership with a project member. Only 12 students were involved in this way, but many of these have gone on to take leadership roles in developing undergraduate research internships and opportunities for further research amongst their peers, to complete PhDs and (in one case) a teaching qualification, to teach their own students in research-based learning ways, and to initiate further academic development initiatives related to undergraduate research.

Working with undergraduate students on projects of this nature challenges academics to treat students as equal partners. In collaboratively implementing projects with their lecturers, students reflected on how they were treated differently in comparison to their classes, thus highlighting the challenges of creating inclusive scholarly communities.

One of our challenges on this project was to ensure that we engaged in democratic discussion (Brookfield and Preskill 1999) with the students working with us (including, e.g. mutual receptiveness, listening, appreciating the contributions of all, and humility). We drew attention to issues of power and authority in our discussions with academics, thus breaking down what Brew (2006, p. 117) describes as an 'academic apartheid', to lay the foundations for students becoming partners in the academic project of the university.

Scholarship

In our program we took the view that undergraduate students should be treated as scholars alongside academics and should be provided with opportunities to experience the scholarships of discovery, integration, engagement, academic citizenship (Boyer 1990) and professionalism (Brew 2006). Through collaborating with us in working groups, in researching practice, in participating in organising and disseminating their research in conferences, and in parliament, students directly participated in these types of scholarship and although the numbers are small, feedback suggests that the effect of participation for many of them was profound as illustrated by this student:

> I gained the understanding that the ability to conduct research is not restricted to academics at university, unlike a common perception that undergraduates only learn coursework and need to wait for higher level study to undertake projects. The great showcase of expertise in differing fields provides a great reason for more undergraduate students to answer their own questions as it is evident that we all are indeed capable of achievement in researching exciting and relevant issues.

An example of how undergraduates created and drove a scholarly culture of research is the establishment of an entirely student-led Undergraduate Research Internship scheme known as MURI. This scheme was established by some of the scholar/ambassadors and others to provide disadvantaged students with the opportunity and support to participate in research with academics, and/or to design their own, or work on other, research.

The peer-led structure of MURI furthers the concept of students as partners in learning. Peer-led group sessions facilitate engagement with research skills and build research interns' confidence to communicate research to each other and the wider community. Students within the MURI program are encouraged to communicate their research to a wider audience. One student commented:

> The most beautiful thing about MURI is the team itself. They provide a learning environment that you feel inclusive, supported and that makes you feel enjoyable to join in and learn. (2014 MURI student)

Another student-led initiative was the undergraduate research student society (MUURSS). Initiated by volunteers from the first Australasian Conference for Undergraduate Research, MUURSS attracted students who had participated in undergraduate research experience programs, and other students with a keen research interest. The Society hosted various events including workshops, stalls during induction weeks, social events and a joint conference with MURI. As ambassadors for undergraduate research across the university, MURI and MUURSS provided pressure groups to lobby for changes to curricula and encouraged undergraduate research within departments.

These initiatives demonstrate the capacity of students to work not just as academics' partners but as initiators and drivers of change. These examples in one institution are mirrored by a growing number of student-led scholarly initiatives in other institutions.

Communities

A number of our initiatives during this period encouraged the development of scholarly communities where students, academics and other staff jointly engaged in promoting undergraduate research. However, there remained negative attitudes of many senior staff to undergraduates engaging in research not just in our university but across the sector. Widespread appreciation of undergraduate research we felt would grow through establishing a community of committed staff and students across the continent of Australasia.

As mentioned above, our project built communities of undergraduate researchers in Australasia and of supervisors of undergraduate research through successive conferences where undergraduates presented their research. These have continued to be held annually. However, their sustainability is dependent upon the willingness of universities to host them and to make funding available for students to attend them.

It is impossible to fully estimate the overall impact of such a complex development program even in one institution let alone across the whole sector. From the evaluations conducted at every stage there is evidence to suggest that the program was important in stimulating a cascade of continuing developments in engaging undergraduate students in research. This is demonstrated in: the ongoing work of former Working Group members developing research-based learning; their continued support for students to attend ACUR conferences; continued student engagement in organisations such as MURI; institutions willing to host the student conferences and, importantly, new strategic institutional research-based learning initiatives.

The idea of universities as inclusive scholarly knowledge-building communities informed the development of a wide-ranging program across seven years of educational development to empower and promote undergraduate research across Australasia. Nevertheless, the program outlined here is just one step in a larger vision. In the next section we bring this work up to date and discuss current developments and future prospects for the development of undergraduate research in Australasia.

2018–2020: Looking to the Future

Approximately 1000 students representing over 40 Australasian universities have been supported by their university to present their research at ACUR events. We know anecdotally that student presenters have made new scientific discoveries; contributed to peer-reviewed journals; given research presentations in disciplinary conferences (including first years being taken for Honours students!); gained prizes and/or funding for their research; taken leadership roles in developing opportunities for further research amongst peers; gone on to complete research degrees, to become academics; and to teach their own students in research-based ways. A full-scale impact evaluation is now in train. However, in line with our vision to develop inclusive scholarly knowledge-building communities our recent concern has been to ensure a lasting legacy of sustainable ongoing development and support of undergraduate research in Australasia.

With this in mind, our task in 2018 was to establish the Australasian Council for Undergraduate Research as an incorporated independent charitable organisation. This consolidated and extended the existing structures to encompass a Constitution as well as financial and reporting responsibilities. ACUR is now administered by a small Executive Group passionate about advancing undergraduate research engagement. It is supported by about 80 people, many holding senior university positions, who represent Australian and New Zealand Universities and some overseas institutions on the ACUR Steering Group. The year 2018 saw the establishment of the ACUR Student Committee which works through social media and in a variety of ways to spread ideas and experiences of undergraduate research amongst students.

The vision for ACUR sees it becoming a peak body for undergraduate research, possibly funding undergraduate research scholarships, providing support for students to attend ACUR conferences and other research events, organising events for supervisors and institutional leaders to disseminate information and ideas, exchanging information about research scholarships available to undergraduates and more fully supporting universities in their efforts to engage their students in research. Since this broader vision is only possible with considerable funding support and, since 2017, ACUR had received no university or government funding, in 2019 membership was introduced. Four categories of membership were established each with their own particular subscription levels, member entitlements and benefits: institutional membership (for Australasian universities); affiliate membership (for other organisations); individual membership, and student membership. (For more information and to join go to acur.org.au/membership. or email: memberships@acur.org.au)

There is much work still to be done. One current initiative to promote and advance the spread of undergraduate research in Australasia is an ACUR Exchange Colloquium to be held in December 2019. This is an opportunity for academics, supervisors, members of the ACUR Steering Group and institutional managers to exchange ideas and hear about good practice in supporting undergraduate research engagement. Our first newsletter, established in 2012 (see above), is now in its 16th issue and has grown to include contributions by student researchers as well as providing information on current developments and upcoming events. Further, a number of Australian universities have joined ACUR as members and we now offer them workshops, resources and consultancy to support them in their efforts to promote undergraduate research engagement.

Conclusion

In reflecting the work that we have done to support and enhance undergraduate research provision, we hope it has been clear that many individuals and universities have contributed to the current state of undergraduate research in Australia. A number of our member universities now have very substantial undergraduate research programs and links with undergraduate research initiatives overseas. Funding for undergraduate research projects remains a particular challenge. However, opportunities exist to align the promotion and advancement of undergraduate research to a range of current and future institutional concerns. A key concern at present, for example, is how to prepare students to be employable through developing graduate attributes for the future of work. Promoting research and inquiry as ways to develop these skills presents an opportunity to further advance undergraduate research experiences for all students.

The model of universities as inclusive scholarly knowledge-building communities presents a bold vision requiring nothing less than a change in institutional culture and as such the work discussed in this chapter has merely scratched the surface. More can always be done but we hope that the account of our work provides a flavour of the challenges we have faced within a small university system. We hope this offers a way of thinking about universities' educational and academic development based on fundamental values of inclusivity, knowledge-building, research, scholarship and the development of university communities, to inspire not just undergraduate research but other important educational initiatives.

Acknowledgements Many people have had a substantial role in the program of development discussed in this chapter. We wish particularly to thank our undergraduate scholars: Evan Jewell and Ademir Hajdarpasic; the student scholar/ambassadors, particularly Aprill Miles, MURI and MUURSS organisers, our collaborating colleagues, ACUR conference convenors, members of the ACUR Steering and Executive Groups, students who have presented their research at ACUR events and their supervisors. Finally, we wish to extend our thanks to the Australian Government Office for Learning and Teaching and to Universities who have provided resources to support the program. This chapter develops the ideas presented in Brew, Mantai and Miles (2019).

References

Australian Government. (2017). *Tertiary education quality and standards agency.* Website https://www.teqsa.gov.au/national-register. Accessed 29 July 2019.

Beckman, M., & Hensel, N. (2009). Making explicit the implicit: Defining undergraduate research. *CUR Quarterly, 4*, 40–44.

Boyer, E. L. (1990). *Scholarship reconsidered: Priorities for the professoriate.* Princeton: Carnegie foundation for the Advancement of Teaching, University of Princeton.

Boyer Commission. (1999). *Reinventing undergraduate education: A blueprint for America's research universities.* Stony Brook: Carnegie Foundation for University Teaching.

Brew, A. (2006). *Research and teaching: beyond the divide.* London: Palgrave Macmillan.

Brew, A. (2010). *Enhancing undergraduate engagement through research and inquiry: National teaching fellowship final report.* Sydney: Australian Learning and Teaching Council. Available http://www.olt.gov.au/resource-enhancing-undergraduate-engagement-research-enquiry-macquarie-2010. Accessed 11 Feb 2015.

Brew, A. (2010). An Australian perspective on undergraduate research. *CUR Quarterly, 31*(1), 37–42.

Brew, A., & Jewell, E. (2012). Enhancing quality learning through experiences of research-based learning: Implications for academic development. *International Journal for Academic Development, 17*(1), 47–58.

Brew, A., & Mantai, L. (2017). Academics' perceptions of the challenges and barriers to implementing research-based experiences for undergraduates. *Teaching in Higher Education, 22*(5), 551–568.

Brew, A., Mantai, L., & Miles, A. (2019). Developing scientific thinking towards inclusive knowledge-building communities: An Australian university case. In M. Murtonen & K. Balloo (Eds.), *Redefining scientific thinking for higher education: Higher-order thinking, evidence-based reasoning and research skills* (pp. 233–253). London: Palgrave Macmillan.

Brookfield, S., & Preskill, S. (1999). *Discussion as a way of teaching: Tools and techniques for university teachers.* Buckingham: Society for Research into Higher Education and the Open University Press.

Hajdarpasic, A., Brew, A., & Popenici, S. (2015). The contribution of academics' engagement in research to undergraduate education. *Studies in Higher Education., 40*(4), 644–657.

Healey, M., & Jenkins, A. (Eds.). (2009). *Developing undergraduate research and inquiry.* York: Higher Education Academy.

Healey, M., Flint, A., & Harrington, K. (2016). Students as partners: Reflections on a conceptual model. *Teaching and Learning Inquiry, 4*(2), 1–13.

Hensel, N. (2012). *Characteristics of excellence in undergraduate research (COEUR).* Washington, DC: Council on Undergraduate Research.

Hutchings, P. (2002). *Ethics of inquiry: Issues in the scholarship of teaching and learning.* Menlo Park: Carnegie Publications.

Jewell, E., & Brew, A. (2010). *Undergraduate research experience programs in Australian universities: National Teaching Fellowship Report.* Sydney: Australian Learning and Teaching Council. Available https://ltr.edu.au/resources/Brew%20A%20Macquarie%20Fellowship%20Undergraduate%20programs%20report%202010.pdf. Accessed 8 Dec 2019.

Karukstis, K. K., & Elgren, T. E. (2007). *Developing and sustaining a research-supportive curriculum: A compendium of successful practices.* Washington, DC: Council on Undergraduate Research.

Kay, J., Dunne, E., & Hutchinson, J. (2010). *Rethinking the values of higher education: Students as change agents.* London: Quality Assurance Agency.

Laursen, S., Hunter, A.-B., Seymour, E., Thiry, H., & Melton, G. (2010). *Undergraduate research in the sciences: Engaging students in real science* (p. 320). New York: Jossey-Bass.

Levy, P., & Petrulis, R. (2012). How do first-year university students experience inquiry and research, and what are the implications for the practice of inquiry-based learning? *Studies in Higher Education, 37*(1), 85–101.

Lopatto, D. (2009). *Science in solution: The impact of undergraduate research on student learning*. Tucson: Research Corporation for Science Advancement.

Matthews, K. E. (2018). Engaging students as participants and partners: An argument for partnership with students in higher education research on student success. *International Journal of Chinese Education, 7*(1), 42–64. https://doi.org/10.1163/22125868-12340089.

Neary, M. (2010). Student as producer: Pedagogy for the Avant Garde. *Learning Exchange, 1*(1). Available http://eprints.lincoln.ac.uk/4186/. Accessed 24 Nov 2016.

Nowotny, H., Scott, P., & Gibbons, M. (2001). *Re-thinking science: Knowledge and the public in an age of uncertainty*. Cambridge: Polity Press.

Popenici, S., & Brew, A. (2013). Reading walls on university corridors: Transitional learning spaces in campus. In M. Vicars & T. McKenna (Eds.), *Discourse, power, and resistance down under: 2* (pp. 145–156). Rotterdam: Sense Publishers.

Smith, P., & Rust, C. (2011). The potential of research-based learning for the creation of truly inclusive academic communities of practice. *Innovations in Education and Teaching International, 48*(2), 115–125.

Spronken-Smith, R., Mirosa, R., & Darrou, M. (2013). 'Learning is an endless journey for anyone': Undergraduate awareness, experiences and perceptions of the research culture in a research-intensive university. *Higher Education Research and Development, 33*(2), 355–371.

Turner, N., Wuetherick, B., & Healey, M. (2008). International perspectives on student awareness, experiences and perceptions of research: Implications for academic developers in implementing research-based teaching and learning. *International Journal for Academic Development, 13*(3), 191–211.

Willison, J., & O'Regan, K. (2007). Commonly known, commonly not known, totally unknown: A framework for students becoming researchers. *Higher Education Research & Development, 26*(4), 393–409.

Zimbardi, K., & Myatt, P. (2014). Embedding undergraduate research experiences within the curriculum: A cross-disciplinary study of the key characteristics guiding implementation. *Studies in Higher Education, 39*(2), 233–250.

CHAPTER 4

Undergraduate Research in the University Curriculum: An Institutional Perspective

Chng Huang Hoon and Wu Siew Mei

Preamble

The National University of Singapore was established in 1905, starting out as a small medical college and evolving steadily over the decades to its current status as a research-intensive institution (see http://nus.edu.sg/about/founded-by-the-community). Like any institution of higher learning, our institution has an established institutional quality assurance process that entails an external review panel undertaking a full-fledged curriculum review. In one such review in the 2004 cycle, an *External Review Panel Validation Report* observed that many reputable universities place value on inducting (junior) undergraduate students in research, and

Chng Huang Hoon (✉)
Department of English Language & Literature,
National University of Singapore, Singapore, Singapore
e-mail: pvochh@nus.edu.sg

Wu Siew Mei
Centre for English Language Communication, National University of Singapore,
Singapore, Singapore
e-mail: elchead@nus.edu.sg

N. H. Hensel, P. Blessinger (eds.), *International Perspectives on Undergraduate Research*,
https://doi.org/10.1007/978-3-030-53559-9_4

did not consider it sufficient that students should be passive consumers of disciplinary knowledge. As an institution that aspires to research and teaching excellence, this gap in our own institutional practice of encouraging undergraduate research triggered the institutional response in 2005 that took the form of a series of actions to provide more opportunities for and recognition of undergraduate research. Among the specific details of this response are (RO.525/05, Outstanding Undergraduate Researcher Prize):

- The introduction of an *Undergraduate Research Opportunities Program* (UROP) in the Science, Engineering and Computing curriculum, in addition to what already exists within faculty curriculum; and
- An introduction of an undergraduate research prize, called the *Outstanding Undergraduate Researcher Prize* (OURP).

Much has happened in the 15 years that have lapsed since this review and plans implementation. UROP, for example, has been implemented in Science, Engineering, and Computing since 2005/2006 and has taken root in other faculties including the Faculty of Arts & Social Sciences (in 2009); and the OURP is an award for best student research projects, and has been open to all NUS students for over a decade.

This chapter presents the research opportunities landscape in our institution by examining the extent in which four of our major faculty curricula have provided avenues for undergraduate student research. These four faculties—Science, Engineering, Computing, and Arts & Social Sciences (FASS)—are chosen because the first three were the original faculties named in the 2005 response to introduce UROP into the curriculum, with FASS coming on board in 2009; and also because today, these four faculties are the four biggest faculties with a current combined enrollment of 18,000 undergraduates (i.e. 64.3% of total current NUS enrollment) and together, they impact a big slice of the NUS student population.

Through a focused email survey and follow-up interviews with the respective curriculum chairs of the above four faculties, we gathered faculty inputs on their curricula and present the different tensions that drive the direction of each of these curriculum. The internal and external factors that have enabled or hindered the infusion of research opportunities within the curriculum are highlighted. These specific cases are offered against the backdrop of the definition of undergraduate student research provided by the Council on Undergraduate Research (n.d.), and contextualized within a consumer–producer framework (Illinois-Urbana Champaign's UR framework, cited in Hensley 2015). We conclude with a reflection of the challenges for the curriculum as these faculties respond to the realities of the nature of the workplace and its demand for "future-ready graduates."

Research, Industry, and the Curriculum

In the "Introduction" to her book, *A Connected Curriculum for Higher* Education, Dilly Fung asks: "can we create better spaces for critical dialogue within and across disciplines?" and to make "better connections between academics, students and 'real world' communities?" (Fung 2017, p. 1) We believe it is important to foster connections between the research that academics do, and the university curriculum and increasingly, also to be connected to the professions. The intellectual capital and research capability available in research-intensive institutions means that they are well placed to make available a wide range of research opportunities for undergraduate students (Howitt et al. 2010; Desai et al. 2008) and to emphasize the scholarly research process in the curriculum. The academic commitment to scholarship is further driven by a firm belief that rigorous disciplinary training has clear derivable benefits, such as cultivating resilience (Lam and He 2019); and gaining skills like design and hypothesis formation, data collection and interpretation, information literacy and the development of professionalism through opportunities for publications and presentations, enhancement of professional credentials (i.e. building a resume), and the development of relationships with mentors and other professionals (Lopatto 2006, cited in Osborn and Karukstis 2009, p. 2).

However, as Harland (2016) puts it, the culture in research-intensive universities where academics as researchers are socialized and incentivized as research beings also often results in research being defined and conducted from faculty perspective (i.e. faculty-led) than from student learning perspective (i.e. students given autonomy to initiate research focus). This strong academic identity that inclines the faculty community toward research, however, is not the only driving force shaping the undergraduate curriculum. As we will show, our undergraduate curriculum faces different internal and external pressures, the latest being pressure from industry stakeholders. Riding in tandem to this is students' pragmatic choice to opt for industry-based training, favoring credited internships and work–study programs over an immersive research pathway, so as to gain an advantage in the world of work. Each faculty curriculum discussed in this chapter has accordingly been reshaped in order to make appropriate responses to these different tensions.

A Consumer–Producer Framework of Undergraduate Research

The Council on Undergraduate Research (CUR) defines undergraduate research as "an inquiry or investigation conducted by an undergraduate student that makes an original intellectual or creative contribution to the discipline." (2019) This definition emphasizes students' engaged learning that takes the form of a research cycle where students, supervised by a faculty member, identify and inquire into research problems through applying specific research methods, ask complex questions, and present their work to peers and instructors, and, in some cases, even publish in undergraduate research journals.

The CUR definition, however, does not specify the conditions under which undergraduate students are enabled to make original contributions to their disciplines (e.g. to what extent the research is student- or faculty-led; the role of the supervisor; or if students function within groups or conduct individual research). In addition, the CUR definition that rests on original contributions may be difficult to achieve for the majority of students, and raises the question if such highly original research is meant for the chosen few (Healey and Jenkins 2009, p. 35). These issues will be discussed toward the end of this chapter after we have presented specific disciplinary cases drawn from interviews with key colleagues overseeing our undergraduate curriculum.

A Continuum of Undergraduate Research (University of Illinois at Urbana-Champaign's Office of Undergraduate Research, cited in Hensley 2015) provides a seven-stage process that is useful in framing the nature of undergraduate research. Undergraduate research develops as a continuum that ranges from student consumption of knowledge on one end (i.e. faculty-led, with little/no student autonomy) to student production of knowledge on the other (i.e. student-defined, with the student initiating the research activity). In between these two end points, students incrementally learn about research and its methodologies, and engage in faculty-led or replicated research toward the end stage of original undergraduate research.

Though students' development of research capabilities may not proceed linearly, the continuum nevertheless characterizes the stages of growth implicit in the research learning process. The research experience initially addresses the learning of research skills and dispositions through what is termed by Healey and Jenkins (2009, p. 6) as research-led, aimed at grasping current research in the discipline (i.e. knowledge consumption) and research-tutored, which exposes students to more active engagement through disciplinary discussions. At these stages, experience with research-orientated situations allows for the development of research skills and techniques, which then lead to research-based experiences that involve the more active undertaking of research and inquiry that can contribute to knowledge building.

Fung (2017) noted that orientation toward what knowledge is affects how research is conceptualized in that discipline. The conceptualization of undergraduate research in disciplines like Engineering and Computing might favour problem solving and product design. In contrast, undergraduate research in the Humanities may lean toward theoretical and ideological debates that push the boundaries of ideas (Aditomo et al. 2013). As such, the actual research tasks that students are involved in as part of their undergraduate research can vary across disciplines, although the underlining processes of engaging students as knowledge consumers and then developing them to be knowledge producers may be shared.

To understand the undergraduate research landscape in the four major faculties—Science, Engineering, Computing, and Arts & Social Sciences—and to gain a deeper understanding of the nature of available research opportunities, we issued an email to the key colleagues overseeing the curriculum in four faculties, requesting for an overview of key research opportunities each has in the curriculum. This was followed by two focused interviews with these curriculum chairs. We also bring to this study our direct experience as academic leaders who have worked in our institution for decades; and the many conversations about the undergraduate curriculum we have had with colleagues from across the university. We discuss these curricular overview and the details of the dominant research opportunities available in each faculty in what follows, keeping both the CUR definition and the undergraduate research continuum frame in view.

The Dual Research-Industry Pathways in the Curriculum

Undergraduate Research/Industry Pathways in the Science Curriculum

Research platforms in Science	*Year 1*	*Year 2*	*Year 3*	*Year 4*	*Remarks*
Independent Study Modules/ ISMs (4 credits)		Elective; eligibility criteria: GPA of 4.5 (except students from University Scholars Program) and above; max of 2 ISMs			
Undergraduate Research Opportunities Program/UROP from 2009 (4–8 credits depending on duration)		Elective; students choose from a list of faculty defined projects			Students moving toward UPIPs
Undergraduate Professional Internship Program/UPIP, from 2015/2016 (4–12 credits depending on duration)		An elective for the majority but compulsory for Food Science students.			
Final Year Project/ FYPs; in existence for decades (12–16 credits depending on duration)				A degree requirement but replaceable by FYIs; taken across two semesters; faculty defined—students choose from a list of available projects	Students moving towards FYIs

(*continued*)

(continued)

Research platforms in Science	*Year 1*	*Year 2*	*Year 3*	*Year 4*	*Remarks*
Final Year Internship (FYI); from 2015/2016 (16 credits full time)				An alternative to FYP for majors like Life Science, Chemistry, Food Science, Environmental Science Data Science; taken across one semester; project defined by faculty-approved companies, co-supervised by faculty (mainly administrative) and industry person	>50% of students opt for this pathway
Lab-based courses; 4 credits, in existence for decades	Fundamental content focus in Years 1 and 2, moving to advanced content in Year 3; and then into more research-oriented work in Year 4; elective in some majors but compulsory for Life Science and Chemistry				

A Science Education is an induction into the scholarly knowledge-building scientific community, with students taught to adopt a scientific attitude toward discovery, with a strong focus on hypothesis formulation and testing. Students embark on their scientific journey of discovery through an immersion in laboratory experimentations from Year 1, with research defined as an activity that embeds the stages of discovery from observation through hypothesis testing, to knowledge generation. The Science undergraduate student in NUS (and elsewhere) is literally thrown immediately into essential lab courses that required students to make observations, test hypotheses, and perform analysis. The centrality of experimentation in the curriculum defined around the process of knowledge discovery may be said to define what it means to be a student of Science.

The Science curriculum provides a fairly wide range of undergraduate research platforms that include lab-based courses for all years, UROP and UPIP in Years 2 and 3; to the final year project in Year 4, and within the last five years, other options like internship are more widely available. All these research options are undertaken by individual students (rather than group, except for Math), and most are faculty-defined though the student can choose or sometimes negotiate the specific topic. Internships (UPIP and FYIs) however, are industry-led, though the Faculty determines which companies to put on their approved list of internship hosts.

Given these available research opportunities, one possible route for a Science student is to pursue a research-focused pathway by accessing up to 16 credits in UROPs and 16 credits for the FYP—this total of 32 credits takes up 20% of the entire Science 4-year curriculum. However, in recent years, this pathway has faced stiff internal and external pressures and competition from the promotion of an industry-focused pathway. Internally and externally, the Science Faculty is under pressure to be responsive to the changes and demands of the workplace, with the central administration speaking a life skills discourse and the employers asking for more "workplace- ready" graduates. The Science student who chooses to privilege industry experience can forego the research path and access up to 16 credits in UPIPs and another 16 credits through the FYI—again making up 20% of the Science degree. The Faculty has reported a trend of "losing" students from the academic/research pathway to the industry-based pathway, with as much as 50–60% of Science majors choosing the internship route. This does not bode well both for faculty who are in need of research students in their labs, and also for a curriculum that aspired to be more integrated and research-based.

In short, in today's push for employability of graduates, and stakeholders' calls for "industry relevance" in the way universities deliver the curriculum, the scholarly commitment to infusing discovery and research, in building a knowledge-based community of scholars, has been somewhat diluted as research opportunities give way to more skills-based teaching. The emphasis on scholarship and community and their intersection with teaching, learning, and research is forced to yield to pragmatic considerations and industry needs.

Undergraduate Research/Industry Pathways in the Engineering Curriculum

Research platforms in Engineering	*Year 1*	*Year 2*	*Year 3*	*Year 4*	*Remarks*
Independent Study Module/ ISM (4 credits)		Elective, typically student-defined			Can be a group project (e.g. in Math)
Undergraduate Research Opportunities Program/UROP (4 credits)		Elective, Individual or group, 2 semesters max, can be student-defined or negotiated or faculty-defined			Only ONE UROP allowed
Final Year Project/FYP (6–12 credits depending on duration)				May be an elective or compulsory depending on the program; can be student-defined or staff-defined	About 10% are student defined projects
Internship (10 credits)			Compulsory, semester-long, industry-driven		Attachment to research institutes possible but <10% in this category
NUS Overseas College (NOC); 6 months to 1 year startup stint (35 credits)		Elective path that is usually undertaken in Year 2 or Year 3			Student numbers increased from 73 to 100 per year; subsumes compulsory internship option
Innovation & Design Program, IDP (48 credits)	A second (smaller) major; an elective pathway				
Co-Op program (started in 2019)	52-week attachment with a relevant host and 16 weeks for the capstone project; an elective program				6 students in the pioneer batch

In the Engineering curriculum, several undergraduate platforms are available for students to gain some exposure to a mixture of research and industry-based experiences. These are the compulsory internship (launched in 2015/2016), UROP (since 2005), and the Innovation & Design Program (about a decade old) in addition to long-standing platforms like ISMs and FYPs. Similar to Computing (see below), while fundamental "deep tech" research is undertaken by faculty members and underpins the work of a professional engineer, there is an applied dimension that focuses on solving real problems presented by the Engineering industry. The Engineering curriculum appears to be geared directly toward producing graduates who are able to meet the holistically predefined outcomes from specific accreditation boards, with 75% of the curriculum aligning with Board requirements. Our interview with the Engineering curriculum chair revealed that this Board specification rests well with the faculty.

In Engineering, research projects are platforms to train students to apply (rather than to discover) Engineering principles to solve "real life problems" so that they can one day be accredited functional engineers. Most Engineering courses have a problem-solving orientation. Technical electives (with a small project tagged on as a course requirement) that teach specific principles and skills occupy prime space in the curriculum. Similarly, UROP, while available from the sophomore year for Engineering students, is an optional pathway, and so are some departments' final year projects (FYPs). Many research projects remain faculty-led (though the specific details may be negotiated) rather than student-initiated (except for ISMs). The relatively low enrollment numbers for some of these research opportunities testify to students' own preference to privilege professional practice rather than scholarly pursuits. Like Science, there is a perception among students that the research path is a more challenging route to pursue, and that there is "no future" in pursuing a scholarly route. The general opinion among colleagues we spoke to over the course of our work in the university is that research lies outside many students' comfort zone and is not favored as a pragmatic choice of a career, given how small Singapore academia is. In this light, scholarship is seen as separate, though embedded, for the most parts, as the academic communities in these disciplines pursue their interests in fundamental research, in knowledge generation and serve as expert practitioners in their industries, even as they train students primarily to become professionals practicing in more problem-driven professional work domains, as members of the professional (as opposed to the scholarly) community.

There exists, however, a number of special program options in the Engineering curriculum. The Innovation and Design Program in Engineering affords multiyear training that can potentially allow highly capable students to innovate and generate new knowledge. Such special programs teach relevant frameworks and methods and counts for one-third of the degree for the 150 students who enroll each year. The work–study Co-Op program is new and has only 6 students in the first cohort undergoing a 68-week stint that exposes them to different pathways. The NUS Overseas College (NOC) route is an entrepreneurial path that emphasizes startup. Engineering reports about 10% of its students choosing NOC in the past year, with gradual increment experienced each year. Together, these multiple pathways are available to students with different capability levels and interest.

Undergraduate Research/Industry Pathways in the Computing Curriculum

Research platforms in Computing	*Year 1*	*Year 2*	*Year 3*	*Year 4*	*Remarks*
Independent Study Module/ISM (4 credits)	Can be taken anytime, but generally from Year 2 onwards, an elective				
Undergraduate Research Opportunities Program/UROP (8 credits)			Elective, across two semesters		About 10 students per semester
Final Year Project/ FYP (12 credits)				Elective, across two semesters	Up to 15MC for Double Degree Program students
Internship (12 credits)			Compulsory; usually undertaken at Year 3		Can be replaced by FYPs if student's GPA is 4.0 and above

(*continued*)

(continued)

Research platforms in Computing	*Year 1*	*Year 2*	*Year 3*	*Year 4*	*Remarks*
NUS Overseas College (NOC); 6 months to 1 year startup stint (between 24 and 48 credits depending on duration)		Elective path that is usually undertaken in Year 2 or Year 3			Over 120 students per semester
Turing Program	4-year elective program for highly capable research inclined students; includes a UROP and FYP component that gives 24 credits.				About 15 students per year.
Co-Op program, in existence since 2017/2018	Up to 64-week attachment to a relevant host				An elective program for Info Security and Business Analytics majors; about 10 students per year

The Computing curriculum is very similar to the Engineering curriculum in many ways, with Final Year Projects/FYPs and internships requirements, elective UROP, and ISMs. A student in Computing can access up to 28 credits in industry-based internships at both faculty and university levels—17.5% of the total curriculum space. In addition to this, a student could also access an NUS Overseas College/NOC to pursue a startup experience, and this could take up to a full year of study abroad (i.e. up to 48 credits). A student pursuing a more research-intensive pathway, however, can also gather 24 credits if he/she is enrolled in the Turing Program, and with a UROP thrown into this mix, could possibly earn up to 32MCs on this path. Like the Engineering program, Computing also allows for a Co-Op pathway of 64 weeks for two of their major program students, namely, Information Security and Business Analytics, though only 10 students per year may take advantage of this route.

Because Computing students are expected to function professionally in the Computing industry, research in Computing may be pure or applied, or both. For example, the Turing Program, aimed at highly capable students is pitched toward pure research but also has an applied dimension.

A visit to the Computing webpage says that the Turing Program encourages students "to pursue fundamental work, to take bold new direction, and to make concrete contributions to the world. This special programme therefore aims to nurture students who aspire to engage in a pure research career in computing." The program description further states: "The Turing Programme is most suitable for students who love to solve *technically challenging* problems and are able to handle both *theoretical* and *practical* work." This special program teaches relevant frameworks and methods and caters to students who meet and maintain a minimum GPA score of 4.0 (out of 5.0) throughout their candidature. About 15 students are trained annually in this program.

What perhaps distinguishes Computing from the other disciplines discussed here is the relative autonomy the faculty has in determining its curriculum. The Computing School does not have an Accreditation Board to align with, and employability for its graduates is not an issue in today's demand for computing and computational expertise. The strength of the market demand for able computing graduates has allowed some room for Computing to dictate its curriculum. The high demand for graduates who can work in coding, programing, and data science is such that students are gravitating not just to the Computing school (with enrollment increased sharply over the years from a few hundred to 1200 in 2019), but also toward the industry, with declining number of students opting to pursue the research pathway that many perceive to be either challenging or as not all that needed for a bright future. The challenge for students opting for the industry path, however, is the struggle to fit what needs to be completed within the semester frame.

Undergraduate Research/Industry Pathways in the Arts & Social Science (FASS) Curriculum

Research platforms in FASS	*Year 1*	*Year 2*	*Year 3*	*Year 4*	*Remarks*
Independent Study Module/ ISM (4 credits)				Elective, student initiated but negotiated with supervisor	Students need to choose between ISM and HT

(*continued*)

(continued)

Research platforms in FASS	*Year 1*	*Year 2*	*Year 3*	*Year 4*	*Remarks*
UROP; in existence since 2009 (4 credits)		Elective, faculty-led (for disciplines like Psychology)			
Honors Thesis/HT (15 credits)				Elective available for those who qualify (3.5–4.0 GPA); 8000–12,000 words; faculty defined topics but student can negotiate in most departments	Students need to choose between an HT or ISM
Internship (4–8 credits)	a) Elective, department-defined internships are aligned with the discipline and hence the most restrictive and therefore less popular b) Elective, Faculty-based internships are most flexible in terms of allowing students to choose an industry of their choice				Students can do more than one internship, gathering 12 credits in some cases

The Faculty of Arts & Social Sciences (FASS) admits a cohort of about 1500 students (roughly 20% of university annual enrollment) annually, of which 300 (or 12% of Faculty enrollment) eventually become Psychology majors, one of the most popular FASS departments (the others being Economics, Sociology, and Communications & New Media). All FASS departments have offered UROP since 2009, though the most recent university data showed Psychology to be the most UROP-active departments, due largely to the department's lab-based nature. All FASS departments share roughly the same range of research platforms. The Honors thesis (HT), for instance, is a faculty fixture though FASS departments vary in their rules relating to which category of students could access this option. In Psychology for example, a GPA of 4.0 (out of 5.0) is needed to access the thesis option whereas other departments may require just 3.5. The independent study module/ISM is an alternative to the HT, and due largely to the shortage of faculty supervisors, an FASS student today has to choose between an ISM or HT. With the range of research-type activities available to students, while an FASS student can take up to 27 credits (almost 17%) in research experience, it is possible to avoid this path totally (except for the minimum 4-credit space for the compulsory lab-based course in Psychology).

In the last decade, FASS has made the internship available as an elective. In August 2019, FASS launched an "FASS 2.0 Industry Tracks" that "offer[s] students the option to complement their Arts and Social Sciences education with industry-relevant training and experiences, whilst pursuing their majors" (fas.nus.edu.sg, accessed 21 November 2019; Teng 2019). The five tracks that were identified after consultations with key internal and external stakeholders encompass major industries that FASS graduates commonly work in, including public administration and banking. Each industry track requires the student to access an industry seminar, capstone career preparation and skills modules, and internship, which totaled 16 credits (i.e. 10% of the curriculum). This provision of an additional embedded pathway within the broader faculty curriculum is a response to the "workplace readiness" discourse circulating both at state and industry levels and to some extent, students demand for internships (a takeup rate of at least 40% of an FASS cohort). The tensions within the curriculum are evident.

Within FASS, research is still largely understood as involving a substantial amount of literature review, some amount of data gathering, quantitative and/or qualitative analysis and interpretation. The output is to extend an argument, or offer a new way of understanding a human phenomenon. But due to the diversity of FASS disciplines, spanning the Humanities to the Social Sciences, the details involved in research may differ. For example, a Humanities subject like English Language has standard research projects (e.g. the Honors thesis) which are largely narrative essays with the necessary literature review and some amount of quantitative and/or qualitative data. This situation is in stark contrast with a Social Science subject like Psychology, where from the get-go, Psychology students are taught research skills in data gathering, hypothesis testing, and analysis. For the most parts, FASS research projects are faculty-led (e.g. Psychology lab-based research) though there are opportunities (like ISMs and HTs in Humanities) that are often student-initiated.

Discussion

It is evident that the four key faculties we investigated—Science, Engineering, Computing and FASS—have a range of research opportunities for undergraduate students in NUS at different points in their candidature. Some of these avenues are available to a select group of students (e.g. the Turing program) and some are much more open (e.g. ISMs). Many of these research opportunities remain individual rather than group projects in spite of their labor-intensiveness. Most projects are faculty- or industry-defined, though students can negotiate the topic focus (i.e. falling around 5 or 6 on the consumer-producer continuum), with relatively more freedom allowed for student-defined topics especially for the independent study module/ISM and thesis platforms (i.e. moving toward students as producer on the consumer–producer continuum). Increasingly, such research avenues are pushed into the elective or alternative space (e.g. the Final Year Project/FYP and the Honors Thesis/HT are two key examples) and some part of curriculum space appears to be taken over by industry-linked projects realized on internship and other program (like FASS 2.0) platforms.

As to the originality criteria set by the Council on Undergraduate definition referred to earlier, generally, undergraduate research is not expected to be highly original but all the curriculum chairs we interviewed agreed that the project has to be "novel" in some way. In most departments, there is an explicit requirement that the student does not simply replicate an existing study, or focus on just a critical literature review. Collecting some new data, or formulating a new hypothesis, or applying a new framework, can qualify for originality. In these faculties, the high bar for originality is reserved for postgraduate students pursuing a Master or a doctorate, where contributions to the discipline become a key factor.

A varied range of research opportunities are available to the undergraduate student in our institution. About 24–32% of faculty curriculum space could be exercised by (at least some groups of) undergraduate students to pursue research. However, these curricula have been facing a challenge to their disciplinary focus on learning through research and inquiry; namely, the push from within and outside the institution toward more skills-based and industry-linked experiences, and the shift in some students' preference for the industry-defined pathway. There is therefore an ongoing tussle between the intellectual emphasis in undergraduate education and the trend toward industry exposure for enhanced graduate employability. All four faculties have been asked in recent years by both the university management and industry partners to infuse more "industry relevance" in the respective curriculum through internships and industry-linked projects because alumni and employers' feedback have indicated that some graduates do not appear to be as "workplace ready" as some employers like them to be.

The university response to external feedback to recalibrate the curriculum in favor of internships has resulted in the birth of two alternate pathways in these four faculties. For example, Science makes available up to 20% of curriculum for either research-based or industry-based learning; and FASS has just launched five industry tracks (10% of curriculum space) in its newly revised curriculum in 2019. Even so, at least in the past few years, the Science faculty has experienced an "exodus" of students, with more than 50% choosing the industry pathway over the research-based pathway. However, our interview with Science revealed that there is some early evidence that this trend may not continue to climb as steeply as at first introduction of the dual pathways. A recent study by Lam and He (2019), for example, showed that while some students perceived the research pathway to be more challenging, there are also students who recognize the benefits of the research pathway that they felt have also equipped them with life skills (that an industry-based learning is predominantly expected to do). Students who have undergone the research pathway in the Lam & He study have in fact rated "resilience" as an attribute they felt they have gained from the rigor of doing disciplinary research.

In their study comparing students' perceptions of the benefits of research-based (RL) and work-integrated (WIL) learning, Lam and He (2019) found that students who undertook the industry pathway have consistently evaluated RL to be not as good as WIL in terms of enhancing verbal communication and professional awareness. The pragmatic nature among many Singapore students may also have a substantial influence on students' choice of industry-linked routes (Choo 2019), as there is the belief, though not fully substantiated, that having multiple internships increases the possibility of a future job with the specific internship host. Added to this is the perceived difficulty of research pursuits, as Lam & He found among their study subjects.

The balance tilting toward the industry-linked curriculum pathway is further boosted by the fact that the government, the university management, alumni, and the Singapore industry have actively taken up the dominant global discourse about the need for "future-ready graduates," who are equipped with life skills (read: nonacademic or beyond academic), who can nimbly function in a future workplace that is shaped by the wider forces of technological advances and the social needs of a first-world society. All these, together with some students' preferences for a more "realistic" dimension to their education, have exerted much pressure in the recent five years on the university curriculum.

What does all this mean for those of us who are committed to a disciplinary research-driven curriculum? What benefits can we articulate to persuade students and other stakeholders that there is still great value in maintaining an intellectual focus in undergraduate education? It is heartening that there are students who have themselves expressed the value of disciplinary training. Our interviews with the curriculum chairs revealed that students regularly observed in their final presentation of their final-year projects and UROP that these experiences have deepened their awareness and increased their understanding of the discipline, connecting for them the theoretical base of the discipline with the application that their projects required them to do. Moreover, student informants in the Lam & He study have evaluated WIL to be not as good as RL in terms of cultivating resilience, and infusing them with disciplinary knowledge and skills. Another finding in this study revealed that RL students perceived the RL path to be more challenging than WIL. For example, "resilience" has the highest mean rating by RL students, suggesting that RL students recognized its importance and had learnt resilience after going through a more intellectually rigorous program.

Interestingly, quite apart from students' own perception of their learning gains on either pathways, Lam & He found that there were no significant differences in perceived learning gains between RL and WIL for employability and most interpersonal skills (except verbal communication, which registered higher in the WIL group). The authors believe there is opportunity for us to shape students' perception of the relative advantage offered by either pathway. In addition, they believe that if we explicitly articulate the value of a more rigorous disciplinary training to students (e.g. that research trains the mind and builds resilience in addition to fuelling disciplinary curiosity), we may be able to counter students' perceptions that may have contributed to the rising popularity in the WIL pathway and the migration of students from RL to WIL.

In addition, Fung has argued that there are ways to embed work-related learning opportunities into the curriculum that allow students to develop work-related knowledge and skills (Fung 2017, pp. 87–88). This kind of embedding is what FASS 2.0 and the mixed pathways offered by Science have done. This may mean allowing for "messier," less academically controlled activities or relinquishing some curriculum space to accommodate the realities of the workplace. There is, however, a need to explicitly guide students to translate and connect their disciplinary learning with their experience in industry and to communicate their training to external stakeholders.

The different kinds of pressures on curriculum aside, faculty mindset toward research, industry-based education, and the curriculum remains a perennial issue for the institution. Our direct experience as academic leaders of curriculum in our institution informs us that many faculty members still largely compartmentalize their undergraduate teaching and their own scholarship, seeing the two as not always aligned, as is evident from the way curriculum designers separate research opportunities from taught courses (except in the special research-intensive program). Furthermore, the issue relating to student autonomy in defining their own research agenda requires faculty members to be prepared to relinquish some amount of control we traditionally exert over the curriculum. As Fung (2018) noted, students need "to develop a strong and confident voice" (p. 86). She further argued that "By learning richly through active enquiry from the beginning to the end of their degree programmes, students engage critically with the kaleidoscope of pictures and voices that surround them and confront the importance and limitations of evidence and 'truth'. In doing so, they not only acquire the knowledge, understandings, and skillful practice they need for the future, they also explore and develop their own identities, places, and voices in the academy, in the professions, and in the world." (p. 86) A radical form of course redesign is therefore needed, which would enable students to "connect academic learning with workplace learning" (Fung 2017). The basic disciplinary training should be provided for all students, whatever pathways they choose, because we believe this cultivation of a habit of mind, of curiosity, rigor, and resilience are important life skills too. What is of crucial importance too in the years ahead is to facilitate the connection to the world of work and granting students the autonomy to define those connections and their journey through the curriculum.

Epilogue

This chapter has outlined the undergraduate research landscape in four main faculties at our institution. We observed that the university curriculum that once enjoyed a high level of autonomy is now increasingly experiencing the pressure from different stakeholders outside the academy. A balance has to be struck in navigating these different tensions. We believe that a new equilibrium will settle in the next five years as all of us in our different capacities adjust to new needs and deliberate over the benefits of the different pathways.

References

Aditomo, A., Goodyear, P., Bliuc, A.-M., & Ellis, R. A. (2013). Inquiry-based learning in higher education: Principal forms, educational objectives, and disciplinary variations. *Studies in Higher Education, 38*(9), 1239–1258.

Choo, D. (2019, December 3). *Singapore's millennials are practical, non-materialistic and care for community: OCBC survey*. Today Online. Retrieved from https://www.todayonline.com/singapore/singapore-millennials-are-practical-non-materialistic-and-care-community-ocbc-survey

Council on Undergraduate Research. (n.d.). Retrieved from https://www.cur.org/who/organization/mission/

Desai, K. V., Gatson, S. N., Stiles, T. W., Stewart, R. H., Laine, G. A., & Christopher, M. Q. (2008). Integrating research and education at research-extensive universities with research-intensive communities. *Advanced Physiology Education, 32*, 136–141.

Fung, D. (2017). *A connected curriculum for higher education*. London: University College London Press.

Fung, D. (2018). Re-thinking the higher education curriculum: Challenges, possibilities, and dreams. In A. Hejnen & R. van der Vaart (Eds.), *Places of engagement: Reflections on higher education in 2040, a global approach* (pp. 82–87). Amsterdam: Amsterdam University Press.

Harland, T. (2016). Teaching to enhance research. *Higher Education Research and Development, 35*(3), 461–472.

Healey, M., & Jenkins, A. (2009). *Developing undergraduate research and inquiry*. Retrieved from https://www.heacademy.ac.uk/knowledge-hub/developing-undergraduate-research-and-inquiry

Hensley, M. K. (2015). A survey of instructional support for undergraduate research programmes. *Libraries and the Academy, 15*(4), 719–762.

Howitt, S., Wilson, A., Wilson, K., & Roberts, P. (2010). 'Please remember we are not all brilliant': Undergraduates' experiences of an elite, research-intensive degree at a research-intensive university. *Higher Education Research & Development, 29*(4), 405–420.

Lam, S. H., & He, C. (2019, November 18). *Research-based versus work-integrated learning programmes: Is one better than the other in training employability skills and attributes?* Paper presented at the higher education campus conference, National University of Singapore.

Lopatto, D. (2006). Undergraduate research as a catalyst for liberal learning. *Peer Review, 8*, 22–25.

Osborn, J. M., & Karukstis, K. K. (2009). The benefits of undergraduate research, scholarship, and creative activity. In M. Boyd & J. Wesemann (Eds.), *Broadening participation in undergraduate research: Fostering excellence and enhancing the impact* (pp. 41–53). Washington, DC: Council on Undergraduate Research.

RO.525/05, Outstanding Undergraduate Researcher Prize, National University of Singapore, 2005.

Teng, A. (2019). *Not your father's arts degree, says NUS faculty dean*. Retrieved from https://www.straitstimes.com/singapore/education/not-your-fathers-arts-degree-says-nus-faculty-dean

CHAPTER 5

Preliminary Perspectives on Undergraduate Research in Egypt

Amani Elshimi

Background

The higher education landscape in Egypt is a massive, bustling and rugged place, yet one holding promise for growth and yield. The Ministry of Higher Education and Scientific Research's (MoHESR) 2018 statistics list 27 public universities (including the Islamic Azhar University, which alone has 26 branches), private universities, and 188 technical, nontechnical and vocational colleges—the later are non-university degree-awarding higher education institutions. The number of undergraduate students enrolled in all departments in 2018 totaled 2,901,209 (*Manẓūmaṭ* 2018). Such high numbers of youth capital promise great potential. Despite various campus challenges—such as over-enrollment, underfunding, and dilapidated facilities in public universities, and soaring tuition costs in private universities—there are multiple examples of undergraduate student engagement and inquiry on Egyptian campuses. Some of these examples are intentionally

A. Elshimi (✉)
Rhetoric and Composition, American University in Cairo, Cairo, Egypt
e-mail: ashimi@aucegypt.edu

N. H. Hensel, P. Blessinger (eds.), *International Perspectives on Undergraduate Research*,
https://doi.org/10.1007/978-3-030-53559-9_5

enabled by supportive university structures, others are driven by students or faculty mentors. This chapter provides a brief overview of multiple university successes, followed by a detailed case analysis of the American University in Cairo (AUC), a private liberal arts institution with an evolving infrastructure for supporting and expanding undergraduate research. These developments are placed, where appropriate, in the context of the relevant Arabic and English published scholarship on undergraduate research.

Undergraduate Research in Egyptian Public Universities

Various conceptions of research exist in the literature, offering different frameworks to understand how universities approach and integrate undergraduate research within their structures. The experience of Egyptian universities in recent years is examined, here, specifically within Angela Brew's model (2003).

In 2019, Egypt articulated national strategic goals for the advancement of higher education and research, outlining a number of achievements and initiatives that focus on the development, modernization, and rank-enhancement of Egyptian universities, and on the strengthening of student competencies and career preparedness, especially in the areas of entrepreneurship and development-oriented creative problem-solving (*Al-Khutta al-Tanfidhiyya* 2019). The years 2017–2019 offered opportunities for engaging young scholars and policy makers together in dialogue and strategic planning. A World Youth Forum was organized in the coastal city of Sharm el Sheikh, both in 2017 and 2018, inviting thousands of young people from over 100 countries to participate in discussion of global issues, largely with a vision to forward the United Nations Sustainable Development Goals. In April 2019, the first Global Forum for Higher Education and Scientific Research (GFHS) was, also, held, with the "intention of conducting insightful action-oriented discussions and attaining wide-ranging perspectives for the purpose of envisioning the next 10 years of Higher Education and Scientific Research in a global, interconnected, and borderless world context" (GFHS 2019). Instituting opportunities to support undergraduate inquiry and experiential learning was part of the discussion.

While there is no articulated definition or national strategy for developing "undergraduate research" pedagogies and structures, the term "action-oriented" may be a keynote, defining not just the discussions on educational reform, but the type of research planned for and expected of university students. The World Youth Forum, the Global Forum for Higher

Education, and the Ministry strategy place emphasis not on "an original intellectual or creative contribution to the discipline," as is the Council on Undergraduate Research definition, but on an original *solution to problems* in the external physical environment. There is barely any reference in the language of the Global Forum agenda and the MoHESR goals to engagement in research in the sense of analyzing the complexity of concepts, questioning theory, understanding the nature of disciplinary knowledge, or participating in the construction of knowledge. The emphasis comes from a business/industrial discourse, employing terms such as "innovation," "industrial revolution," "emerging technologies," "global challenges," "digitalization," and "leadership."

These opposing conceptions of research are neatly interpreted by Angela Brew's (2003) model of research. Brew offers an interesting categorization of research, mapping the relationship between research and teaching in quadrants, based on whether the priority of the research is the external product or internal process, and whether the researcher is the focus of the process or not.

Table 5.1 illustrates Brew's model.

Brew explains each of the "views," concluded from a qualitative study that analyzed interview responses from 57 Australian educators, as follows (2003, 6–7):

- Within the external dimension, expectations are "external to the process of doing research." The Trading view prioritizes products that focus on the advancement of the researcher, such as grants and published papers. The Domino view prioritizes products that "push the frontiers of knowledge" such as answers or solutions to complex issues. The researcher is not the focus.
- Within the internal dimension, expectations "look inward." The Journey view prioritizes the researcher's "transformation" and

Table 5.1 Brew (2003, 6)

Research is oriented towards	Research aims to	The researcher is present to, or the focus of, awareness	The researcher is absent from, or incidental to, awareness
External products	Produce an outcome	Trading view	Domino view
Internal processes	Understand	Journey view	Layer view

"journey of discovery." The Layer view prioritizes the meaning-making, and the creation of "reliable, systematic information."

The action-oriented philosophy, therefore, that guides undergraduate research engagement in most Egyptian universities, may be interpreted within Brew's external *Domino* view, where the focus is on solving problems and addressing local and global issues. Researcher development is not the focus; nor is the *process* of research and critical inquiry. The important outcome is the significance of the *product*. This is not a new vision. Even historically, despite dire conditions of the educational and physical infrastructure for research in public universities, it is undergraduates in the applied disciplines of the natural sciences, engineering and architecture, computer science, agriculture, dentistry, and media studies, amongst others, who have long been required to submit a research-based graduation project, offering a new mechanical technique, solution, or innovation. Each of these disciplines normally offers a capstone research methods course, where groups of students engage in their first and/or most significant experience with hands-on research, mostly mentored by teaching assistants (TAs). Most students work with meager self-funded (usually parent-funded) resources, scarce essential materials and equipment, often oversized groups of varied competencies, and limited access to key information sources. Despite these limitations, the students persevere and work diligently, building autonomy and agency, often winning global competitions for scientific innovations, computer applications, robotics, and other technical and business advances. Examples of such competitions include Alexandria University students winning first place at the European Union Competition on Entrepreneurship Awareness (Eldeeb 2019); Cairo University students ranking highest in the first phase of the Design/Build/Fly competition of the American Institute of Aeronautics and Astronautics (*Cairo University Aerospace* 2018); and Mansoura University geology students participating in the discovery of the skeleton of a dinosaur, which the international research team termed *Mansourasaurus*, in the Egyptian Western Desert (Sallam et al. 2018).

With the new Ministerial strategic plan, universities now take an active role in fostering and supporting undergraduate research. The bigger universities, such as Cairo University, Ain Shams, Alexandria, Tanta, Suez Canal, Assiut, and others, offer funding for research and student travel, as needed. Some, including the Ministry of Defense Military Technical College, organize an annual undergraduate research conference. A faculty member from each of three public universities—Mansoura, Beni Suef, and Zagazig—was interviewed for further detail on the institutional undergraduate research support effort of their respective universities.

Located in the Dakahlia Nile Delta, Mansoura University, established in 1972, and currently hosting 165,000 undergraduates (Mansoura University 2019) is an example of a university that has adopted an intentional strategy for "developing the student's personality to be capable of innovating, challenging, self-learning, working in a team and competing regionally and globally" (Strategic goals 2012). It recently established a Student Research and Innovation Office (SRIO). Housed under the Vice President of Education and Student Affairs, the SRIO offers partial funding for student research projects, and organizes discipline-specific student conferences, mostly in the pharmacology and engineering departments, in addition to an annual conference for student initiatives, co-organized by the students. Once again, most projects are graduation capstones in the technical disciplines, geared toward environmental, science, and business solutions. The office also facilitates supplementary funding for student research through the Technology Innovation Commercialization Office (TICO). This is an intellectual property and technology transfer office, established in cooperation with the Egyptian Academy of Scientific Research.

Beni Suef University, home to 50,000 undergraduate students (Beni Suef 2019), adopts a different undergraduate research approach, placing student research within the purview of community engagement. Research is perceived as a civic skill, and developed for the purpose of building the local community and environment. Student–faculty research partnerships are developed within the science departments, not through a centralized office across the university. All student research activity is overseen by the Vice President of Community Engagement and Environmental Development. Beni Suef University, also, hosts an Institute for Small and Medium Enterprise, which engages both faculty and students. In an interview with a student mentor, Dr. Mostafa Ragab Abdel Wahab, lecturer and researcher in the College of Sciences, cited a number of articles that were the product of mentored community-based research, and coauthored by undergraduate researchers alongside faculty. Adopting the external Domino view of research (Brew 2003), the Beni Suef model further narrows its target beneficiary of student research. Students work to address and directly serve community needs through their research and service, fulfilling the "public purpose" of undergraduate education (Munck et al.

2014, 1). However, unlike the Western discourse on community "equal partnership" (29), which depicts a business-like reciprocal relationship between two entities, the Beni Suef student researchers identify with community. The environment surrounding the university and receiving the "product" of their research is their home, their families, and neighborhoods. The relationship is not one of "partnership" but of organic belonging.

Other less-funded universities, such as Zagazig University, depend on faculty-initiated projects and partnerships. Assistant Professor of Industrial Engineering and Management, Noha Mostafa, reports that students have participated, through a partnership with the Maʿan Arab University Alliance for Civic Engagement (Maʿan)—a university network which is housed by the American University in Cairo—in a research competition, presenting solutions to corporate and industry-posed problems, and winning opportunities and sponsorship for implementation. Some students have also published in undergraduate conference proceedings, managed by the Military Technical College.

In the realm of private universities, three new higher education institutions stand out as being research-directed at the undergraduate level—Nile University, the Egypt-Japan University for Science and Technology (E-JUST), and Zewail City of Science and Technology. Nile University describes and positions itself as "Egypt's Research University." It offers bachelors programs only in business and applied sciences, articulating a research strategy that addresses key strategic priorities in Egypt—healthcare, agriculture and crops, traffic and vehicles, energy and water, software and communications, social applications, and innovation, entrepreneurship, and competitiveness. Its strategy for undergraduate education is articulated as follows: "Students have to hone their skills, search for knowledge and find it for themselves, their role is to question conventional wisdom, investigate it, challenge it and try to improve on it. Faculty's role at NU is to guide, coach and support students in their search for knowledge" (Nile University 2019). E-JUST's research vision emphasizes "introducing effective and efficient research output that is relevant to global trends and meets national priorities" (2019).

In a similar vein, Zewail City of Science and Technology, described as a "National Project for Scientific Renaissance," invests in research and technology that address "strategic challenges on the national and international level. [It aims] at providing new inventions that will contribute effectively in developing societies and enhancing economies" (University Vision

2015). Zewail City University offers undergraduate programs only in Engineering and Science, and administers a "STEM" entrance examination to its applicants. Students work on interdisciplinary hands-on projects, linked to industry, and take a course on intellectual property, technology transfer, and commercialization. The entire pedagogical strategy for undergraduate education is based on immersion in the research process and engaging in problem-oriented scientific inquiry.

It is interesting to observe that while E-JUST and Zewail City embrace the Domino view in Angela Brew's (2003) research framework, focusing on "outputs," "national priorities," "strategic challenges," and "inventions," Nile University adopts the Layer view, focusing on the internal growth of the student researchers, using phrases that express the development of research skills and meaning-making—"hone their skills," "question conventional wisdom," "investigate," "challenge," and "improve." All three universities have formed a coalition, together with the American University in Cairo, known as CUREE—the Coalition for University Research Excellence in Egypt (El Gendy 2019). The coalition seeks to advance and lead faculty and student research achievements.

Egyptian universities, then, especially in the technical fields, are shifting from a knowledge-based to an outcomes-based, or more accurately, a product-based higher education. The new vision perceives the student as a productive agent of change, rather than an intellectual, discerning scholar, engaged in theory analysis and conceptual critique, and asserting authorship (Grobman 2009). It is not "contribution to the discipline" (Council on Undergraduate Research) that is the key characterizing factor of student research, but "technical solution," "entrepreneurial initiative," or "contribution to (global) development." This approach is largely driven by young lecturers, assistant and associate professors who are inspired, inspiring, and passionate about supporting student learning and empowerment. The students develop skills in problem analysis, "original" solutions, and innovation. This is in line with undergraduate experiences in other contexts, where economic development is the primary driving force for student research and workplace preparation (Van Galen et al. 2015).

Undergraduate Research at the American University in Cairo

The American University in Cairo (AUC) is a private, nonprofit institution, accredited by both the American Middle States Commission for Higher Education and the Egyptian National Authority for Quality Assurance and Accreditation in Education (NAQAAE). The 2019 figures show that AUC served a population of 5474 undergraduates, as well as 979 graduate students, bringing an American liberal arts philosophy of education and core curriculum to a 95% Egyptian cohort of students, at a faculty-to-student ratio of 1:11. The case of the AUC is presented here, given its intentional and strategy-driven approach to undergraduate research, based on the American model of undergraduate research as a high-impact practice. The definition of high-impact practices adopted by the university comes from the Association of American Colleges and Universities (AAC&U), specifically the work by George Kuh (2008). Kuh outlines six conditions for high-impact "unusually" effective educational practices: they require "considerable" time and effort; they immerse students in activities where they have to interact with faculty in "substantive" ways; they promote diversity; they provide rich and frequent feedback to students; they enable learning outside of the classroom; and they are transformative, often life-changing (14–17). AUC purposefully designs student research experiences to achieve these conditions.

In ways much similar to the Egyptian universities, AUC's undergraduate students in the technical fields of the sciences and engineering have always had a thesis exit requirement, in contrast to their business and humanities/social sciences counterparts. Course-embedded research, however, has been inherent in the humanities and core general education curricula, with instruction of basic research writing mandatory in the first year writing courses. These include skills such as the formulation of a debatable research question, identification of scholarly sources, critical analysis of literature, becoming aware of the audience and the rhetorical situation, developing a sound argument with logical organization, adopting effective strategies of persuasion, and documenting sources with integrity. In the second, third, and fourth years, many of the disciplines, though not all, offer a course or more in specialized research methods. Every student on campus, also, regardless of discipline, is required to fulfill two capstone courses—one in the department of their major, and one outside the major. The learning outcomes always include a research-intensive capstone project. These are the curricular, credit-based offerings.

Co-curricular research activity began sporadically, with individual faculty members or student organizations organizing public presentation events. In 2004, two faculty members in Rhetoric and Composition, organized the first institution-wide Undergraduate Research Conference, under the theme of "Reform in Egypt: Opportunities and Challenges." The students participated in the conceptualization, organization, marketing, and recruitment for the conference, inviting four distinguished keynote speakers, both AUC faculty and invited guests. Over 50 undergraduate participants presented engaged, well-supported and nuanced research. The conference was hugely celebrated on campus, by both students and senior administration. The organizers collected the student papers, reviewed and edited them, and posted them on the departmental webpage, creating the beginnings of an online publication.

In the following year, 2005, the presentation event developed into the annual conference for Excellence in Undergraduate Research, Entrepreneurship and Creative Achievement (EURECA). A series of faculty-facilitated workshops were given on campus, ranging from "How to Write an Abstract" to "How to Give a Presentation." With the approval of a committee of reviewers, more than 70 students from various disciplines presented their research work in either oral and poster formats, and the ensuing publication officially became URJe—The Undergraduate Research Journal. The idea of student scholarship was born. URJe is now a registered open access publication, hosted on Open Journals System (OJS), and listed in the CUR Undergraduate Journal Catalog.

Over the following years, the conference grew to integrate applicants from other Egyptian universities, and, on a small scale, international applicants. New activities were also introduced to showcase all creative, research-based projects across the curriculum, and to integrate students starting from the pre-credit language instruction courses, through the first-year mandatory research-writing courses, and to the final graduating seniors' thesis-level work. The following became the subcomponents of EURECA:

- Research Excellence Across the Disciplines—oral presentation panels for students in the majors
- First-Year Research Experience (FYRE)—oral presentation panels for freshman students

- ELI Explorers—oral presentation panels for pre-credit language students
- Undergraduate Poster Competition—research poster presentations for undergraduates at all levels
- Creatopia—presentations of original creative works, including:

  - Creative writing, script-writing, spoken word poetry
  - Short film, documentary, podcasts
  - Visual artwork and photography
  - Graphic design
  - Cartooning and graphic novels
  - Creative game design
  - Musical compositions
  - Digital Liberal Arts projects

- Entrepreneurship Expo—for undergraduate business ideas and startups
- Architectural Design Show—for undergraduate design innovations

It took almost eight years for the university to finally institutionalize co-curricular undergraduate research. A new Academy of Liberal Arts (ALA) was established in 2012, to house the general education core curriculum, the non-degree awarding departments of English Language Instruction, Arabic Language Instruction, and Rhetoric and Composition, and the high-impact academic support programs. A place (and space) for an Undergraduate Research Program was, thus, created, alongside the sister programs of Academic Community Engagement, the Common Reading Experience, and the university Writing and Communications Center. The director of Undergraduate Research, now reporting to the Dean of ALA, became one of the two volunteer faculty members who originally organized the student conference in 2004, with a half-release of teaching time to run the program.

The new Undergraduate Research Program articulated a mission to "Institutionalize, support and expand opportunities for undergraduate research and creative achievement; and to nurture amongst the academic community, across the disciplines, a culture of research and development, and the drive to advance the liberal education outcomes of undergraduate inquiry and critical and creative competence" (Undergraduate Research Program). The goals became the following:

- Develop the program, on an ongoing basis, building on the advisory council of key stakeholders
- Promote undergraduate research and creative works through activities, periodic events, and ongoing services to students, graduate fellows and academic faculty members
- Document undergraduate written, oral, and visual outcomes, faculty pedagogical experiences, and graduate fellows' mentoring reflections
- Reward exemplary performance and provide grants to support undergraduate research
- Evaluate program outcomes through continuous research and assessment

Alongside the annual conference, which in 2019 showcased the research and creativity work of over 350 undergraduates, the journal publication, and training workshops, the Undergraduate Research Program began offering travel grants to student presenters at academic conferences, and summer research internship grants to students accepted at research internships abroad. In the first year, the funding came from the Academy of Liberal Arts, but in the third year of the program, 2014, a new budget center was created with $30,000. After an initial dip in funding, due to the inflation of the Egyptian Pound, the program budget was increased in 2017 to $56,000, with budget lines for student grants, publications, professional development travel for the director, and event hospitality. In 2018, a new activity was added to the mandate of the program—an institution-wide writing competition in five different genres of writing: short story, the spoken word, graphic novel, scriptwriting, and podcasting. Two faculty members were appointed to manage the competition, and a $100,000 budget was allocated for marketing, workshop tutor fees, and generous travel awards to writing camps for the winners in each of the five categories. The director and administrative assistant of Undergraduate Research were responsible for budget disbursement and monitoring. On a campus of a largely English as a Second-Language body of undergraduates, the Writing Competition was seen as a vehicle for enhancing the writing ability of students. The budget increase, though, did not extend to undergraduate research activity.

The challenges, therefore, remained the same—understaffing, poor marketing, sometimes faltering technical platforms, and, ultimately, insufficient funding. For a program whose scope was institution-wide, a more robust infrastructure was needed. The budget, though increased, still could not cover the full costs of students accepted at prestigious conferences, such as the Institute of Electrical and Electronics Engineers (IEEE) and the Middle East Studies Association (MESA) conferences, and research internships at universities abroad, including Harvard, MIT, Princeton, and Leeds, amongst others. The Travel Grant for conferences or competitions, covered \$1000, and the Research Internship Grant covered \$1500 of expenditures. For students traveling to the United States, Canada, Australia, and other distant and expensive venues, travel and accommodation costs, especially for an extended one-and-a-half months or more for a summer internship, far exceeded the awarded grant. The result was that student grant awardees constantly sought supplementary funding from various offices on campus. Eventually, a shared google spreadsheet was created to monitor cost-sharing and centralize disbursement. This and other challenges were effectively addressed through forming strategic partnerships with other units on campus to expand and diversify UR activities, share the cost and workload, as well as align with the strategic goals of the university. When, in 2018, UR activities were aligned with the strategic initiatives of the Associate Provost for Research, Innovation and Creativity, the budget escalated the following year a full \$200,000. The undergraduate support grants, therefore, were adjusted to adequately cover travel costs, raising travel grants to \$1800, and research internship grants to \$4000. Grant types were increased to cover competitions, community-based research, and requests for research resources (Undergraduate Support Grants). The UR Program now, also, documents the work of students in a Newsletter for Undergraduate Research, and promotes a summer boot camp on "Writing the Research Proposal," targeting undergraduates from public universities in Egypt.

Strategic Partnerships

The UR program is, incrementally, creating a culture of scholarship among the undergraduate community and the faculty. Through carefully conceived strategic partnerships with campus entities, the program has been able to widen its scope and increase its impact with low or no additional cost or time. Some important gains that have been achieved through collaboration with significant campus partners are:

- *UR Partnership with the Core Curriculum*

  Working with the Core Curriculum Advisory Committee, the UR Program has ensured that the university core curriculum has made explicit learning outcomes that highlight research skills at every developmental level. These include critical thinking and reading, written communication, and information literacy at Freshman and Secondary level, and interdisciplinary and multicultural competencies at capstone level. This learning outcome reads as follows: "Students will engage in inquiry by integrating knowledge drawn from various disciplines and perspectives to address real-world problems and demonstrate a more nuanced understanding of different aspects of local, regional and global issues" (Core Curriculum, AUC). While not strictly based on the "Connected Curriculum" Framework, offered by Dilly Fung (2016), the new core learning outcomes build on Gadamer's (2004 cited in Fung 2016) underlying philosophy of student development. Fung describes the philosophy as one "enabling students to develop themselves, both individually and in communities, through dialogue and through active, critical engagement within and across subject fields" (31). The Core learning outcomes, thus, ensure that research skills are developed in increasing levels of complexity, across courses, and throughout the progression of the core curriculum.
- *UR Partnership with the Department of Rhetoric and Composition*

  To acquire adequate staffing, and to reduce the teaching–admin workload, the director of the Undergraduate Research Program, herself a faculty member in the department of Rhetoric and Composition, developed a RHET-designation capstone course that aimed to offer on-the-job research and writing experience. Students who enroll in the course (capped at 12 students) may be placed in different workplaces—businesses, nonprofits, publishers, or university offices—to participate, as nonpaid interns, in research activity and professional writing. When the director of UR teaches the course, all 12 students enrolled in the course intern with the Undergraduate Research Program. This is a win-win situation, with students gaining credit-bearing job experience, and the director merging teaching/admin time, and gaining the fruit of student labor. The students participate in program evaluation and research about Undergraduate Research, review grant and conference evaluations, participate in journal editing, develop promotional material for the program, and manage the newsletter.

- *UR Partnership with the AUC Career Center*

  A partnership with the AUC Career Center helped create new opportunities for undergraduate researchers. With a budget to fund work–study students, the Career Center was able to create a new category of *research* work–study offerings, providing paid positions for junior and senior student researchers interning with department faculty and research centers on campus.
- *UR Partnership with the Library*

  Partnering with the library boosted the Undergraduate Research Journal (URJe) profile, and expanded the panel of reviewers and editors. The library upgraded URJe into an ISSN-registered open access journal on OJS.
- *UR Partnership with AUC's Center for Learning and Teaching*

  After initially offering faculty a number of UR-focused workshops that were barely attended, the UR Program partnered with the university's Center for Learning and Teaching (CLT). Given that the CLT offers certification for faculty attending professional development institutes, which are then reported on the Annual Faculty Report, the UR Program employed this incentive, offering a semesterly institute on "Integrating Undergraduate Research in and beyond the curriculum." The number of faculty attendees multiplied, with almost 25 participants per session. Six of these faculty members, in 2018, continued to seek advisory on course-embedded undergraduate research, intentionally restructured their courses to embed research outcomes, and became informed advocates for undergraduate research.
- *UR Partnership with the Associate Provost for Research, Innovation, and Creativity*

  In 2017, the Provost started an institution-wide event—the Research and Creativity Convention—to showcase AUC's intellectual output: faculty, graduate students', and undergraduate students'. The EURECA conference was invited to become a key component of this event. The shift accrued many gains, most important of which were marketing and funding. EURECA gained visibility at the senior administrative level, and was communicated across a wider scope, including the Board of Trustees, the university's New York office, the alumni network, and the AUC community at large. The marketing strategy and costs were monitored and covered by the office of the Associate Provost for Research, Innovation, and

Creativity. The costs of promotional material, keynote speakers, and awards were also covered. Finally, a new award—the Grant Award for the Disciplines—was introduced. The awarded $25,000 goes to a department that integrates UR in its mission statement, creates curricular and co-curricular opportunities to advance and support undergraduate research, and recognizes exemplary faculty mentors and advisors.

- *UR Partnership with the Office of the Provost*

  Working with the Provost's Office to enhance opportunities for students working on faculty research, a supplementary incentive fund of $1500 was added to the $10,000 Faculty Research Grant, and made available to faculty who engaged undergraduate research trainees in their research work. The numbers of student beneficiaries are still quite small, ranging between 8 and 12 students per grant cycle.
- *UR Partnership with the University's Life Mentorship Program*

  The Life Mentorship Program supports faculty members in serving as academic and entrepreneurship mentors to undergraduate students. A conversation with the program resulted in the expansion of the mentorship scope to include undergraduate research as well. Faculty research mentors, therefore, would receive an orientation to the philosophies and expectations for undergraduate research, and periodically meet to reflect on experiences, challenges, needs, and achievements. The Mentorship Program will award highly accomplished faculty, based on deliverables, such as coauthored papers, student presentations at international conferences, or student publications.
- *UR Partnership with AUC's Scholarship Office*

  A partnership with the scholarships office considered a redescription of a scholarship program—Tomorrow's Leaders Scholarship Program, which impacts 12 students a year. The funding source requires that students engage in community work, and implement a community-based project. Integration with the UR Program resulted in a required research publication that is based on the project. Students present their research at the annual EURECA conference. Other scholarships are similarly being revisited, perhaps, to embed a research requirement for graduation.
- *UR Partnership with Student Organizations*

> The best spokespeople for the services of the Undergraduate Research Program are students. The Student Union (SU), with its huge database of students and media outlets helps promote the activities and grant opportunities offered by the UR Program. The First Year Program (FYP), a student organization that supports freshman students, helps organize an-end-of-semester presentation and award event for students presenting research posters for a First Year Research Experience (FYRE) competition. This year, 54 students participated in the competition.

AUC is unique in Egypt in its actively evolving vision to ingrain undergraduate research within the whole student experience—curricular, co-curricular, from first year to capstone, student-led and faculty-led, paid, credit-based, and voluntary, in the humanities and sciences, applied and theoretical disciplinary and interdisciplinary contexts. The funding, staffing, and infrastructural challenges help inspire conversation at the administrative level, and almost always culminate in heightened awareness and refined strategy for undergraduate research. Still, there is a gap between the administrative direction, and the deeply anchored beliefs within traditionally taught programs, which perceive undergraduate research and scholarship as secondary to the teaching mission of the faculty. The process of implanting a culture and shared vision in the departments will take time before undergraduate research becomes an institution-wide strategy.

Conclusions

The Egyptian undergraduate research experience is diverse. This brief overview surveys the strategy, infrastructure, activities, and challenges across public and private universities in Egypt, with a focus on the case of the American University in Egypt. The terrain has huge potential for expansion and development. Most universities explored in this study have yet to conceptualize a vision and strategy for integrating undergraduate research practices into both the curriculum and co-curriculum, describe the guidelines that govern adoption of this student-centered pedagogy, and create recognition systems for both the students and faculty mentors. Universities need to work on the spatial capacities and technology infrastructure—the labs, studios, funding, and institutional industrial and community partnerships. All need to invest in creating the culture on campus and building a shared vision amongst faculty and departments. The vision, so far, is not consensual. It may be top-down, with senior administration adopting the national strategic goals for higher education and research, or bottom-up, with a handful of passionate faculty members, engaging students in their own research, and embedding student research in their classes. Ultimately, the students themselves need to identify with their role as scholars, taking ownership of the learning process, and developing a "scholar" persona—a major shift of identity that requires a change of attitude, responsibilities, and expectations. These are some preliminary perspectives on the landscape of undergraduate research at Egyptian universities.

Acknowledgments I would like to thank the following faculty for their generous contributions: Samah Elsaid, Assistant Lecturer of English, Mansoura University; Noha Mostafa, Assistant Professor of Industrial Engineering and Management, Zagazig University; and Mostafa Abdel Wahab, Lecturer and Researcher at the Faculty of Sciences, Beni Suef University.

References

About GFHS. (2019). *Global forum for higher education and scientific research.* Retrieved June 20, 2019 from http://www.gfhsegypt.com/about-gfhs/

Al-khutta al-tanfidhiyya li-wazaarat al-taʿleem al-ʿaali li-ʿaam 2019 [The Ministry of Higher Education 2019 executive plan]. (2019). *The ministry of higher education and scientific research portal.* Retrieved June 20, 2019 from http://portal.mohesr.gov.eg/ar-eg/Documents/plan-mohe-2019.pdf

Beni Suef. (2019). *Times higher education.* Retrieved from https://www.timeshighereducation.com/world-university-rankings/beni-suef-university

Brew, A. (2003). Teaching and research: New relationships and their implications for inquiry-based teaching and learning in higher education. *Higher Education Research & Development, 22*(1), 3–18. https://doi.org/10.1080/0729436032000056571.

Cairo University Aerospace Students Garner First Place in AIAA Competition. (2018, November 26). *Egypt independent.* Retrieved from https://ww.egypt-independent.com/cairo-university-aerospace-students-garner-first-place-in-aiaa-competition/

Core Curriculum Learning Outcomes. The American University in Cairo. Retrieved from https://www.aucegypt.edu/academics/core-curriculum/core-curriculum-learning-outcomes

Council on Undergraduate Research. Mission. Retrieved from https://www.cur.org/who/organization/mission/

E-JUST Research Vision and Status. (2019). Egypt-Japan University for Science and Technology. Retrieved from https://ejust.edu.eg/e-just-research-vision-and-status/

El Gendy, N. (2019). New scholarships, University Research Coalition Launched at Chinese-African Forum. *The American University in Cairo*. Retrieved from https://www.aucegypt.edu/news/new-scholarships-joint-research-chinese-african-higher-education-forum

Eldeeb, I. (2019, October 19). Alexandria University scoops EU competition on entrepreneurship awareness. *Sada El Balad*. Retrieved from https://see.news/alexandria-univ-ranked-1st-in-eu-competition-on-raising-awareness-about-entrepreneurship/

EURECA Conference. The American University in Cairo. Retrieved from https://www.aucegypt.edu/research/undergraduate-research/eureca-conference

Fung, D. (2016). Engaging students with research through a connected curriculum: An innovative institutional approach. *Council on Undergraduate Research Quarterly, 37*(2), 30–35.

Gadamer, H. G. (2004). *Truth and method* (2nd rev. ed., J. W. Marshall, Trans.). London: Continuum. Cited in Fung, D. (2016). Engaging students with research through a connected curriculum: An innovative institutional approach. *Council on Undergraduate Research Quarterly, 37*(2), 30–35.

Grobman, L. (2009). The student scholar: (Re)negotiating authorship and authority. *College Composition and Communication, 61*(1), 175–196. http://www.ncte.org/library/NCTEFiles/Resources/Journals/CCC/0611-sep09/CCC0611Student.pdf

Kuh, G. D. (2008). *High-impact educational practices: What they are, who has access to them, and why they matter*. Washington, DC: Association of American Colleges and Universities.

Maʿan Overview. The American University in Cairo. Retrieved from http://schools.aucegypt.edu/academics/maan/Pages/default.aspx

Mansoura University. (2019). *Top universities*. Retrieved from https://www.topuniversities.com/universities/mansoura-university/undergrad

Manẓūmaṭ al-taʿleem al-ʿali fi Misr: Naẓra ʿamma [The organization of higher education in Egypt: A general overview]. (2018). *The ministry of higher education and scientific research portal*. Retrieved June 20, 2019 from http://portal.mohesr.gov.eg/ar-eg/Documents/

Munck, R., McIlrath, L., Hall, B., & Tandon, R. (2014). *Higher education and community-based research: Creating a global vision*. Palgrave Macmillan. Retrieved from https://www.researchgate.net/publication/309731294_Higher_Education_and_Community_Based_Research_Creating_a_Global_Vision

Nile University Undergraduate Programs. (2019). *Nile University*. Retrieved from http://nu.edu.eg/undergraduate-programs/

Sallam, H. M., Gorscak, E., O'Connor, P. M., et al. (2018). New Egyptian sauropod reveals late cretaceous dinosaur dispersal between Europe and Africa. *Nature, Ecology and Evolution, 2*, 445–451. https://doi.org/10.1038/s41559-017-0455-5.

Strategic goals. (2012). *Mansoura University*. Retrieved from http://www.mans.edu.eg/en/about/info/strategic-goals

Undergraduate Journal Catalog. *The Council on Undergraduate Research*. Retrieved from https://www.cur.org/resources/students/journals/catalog/

Undergraduate Research Program. The American University in Cairo. Retrieved from https://www.aucegypt.edu/research/undergraduate-research/undergraduate-research-program

Undergraduate Support Grants. The American University in Cairo. Retrieved from https://www.aucegypt.edu/research/undergraduate-research/support-grants

University Vision. (2015). Zewail City of Science and Technology. Retrieved June 2019 from https://www.zewailcity.edu.eg/main/content.php?lang=en&alias=university_vision

URJe – The Undergraduate Research Journal. *The American University in Cairo*. ISSN: 2356–8976. Retrieved from https://urje.aucegypt.edu/index.php/AUC

Van Galen, D., Schneider-Rebozo, L., Havholm, K., & Andrews, K. (2015). Undergraduate research and economic development: A systems approach in Wisconsin. In M. Malachowski, J. M. Osborn, K. K. Karukstis, & E. Ambos (Eds.), *Enhancing and expanding undergraduate research: A systems approach. New Directions for Higher Education, 169*, Spring, 39–49. Jossey-Bass. Wiley Periodicals.

CHAPTER 6

Undergraduate Research in Japan: Beyond the Dichotomy of Product-Oriented and Process-Oriented Approaches

Rintaro Imafuku

Introduction

Undergraduate research (UR) is a high-impact educational practice in higher education and beneficial for students (Beckman and Hensel 2009; Kuh 2008). Research can be a process of learning or discovery (Brew and Boud 1995). In comparison to many other countries, a student development and process-centered perception of research is relatively new to undergraduate education in Japan. This is because research has traditionally been viewed from a content-oriented or product-centered perspective in Japan. The definition of UR varies across countries, institutions, departments, and individual faculty members. This variation within UR components and practices can be viewed along the following continua: process-centered versus product-centered; student-initiated versus faculty-initiated; all students

R. Imafuku (✉)
Medical Education Development Center, Gifu University, Gifu, Japan
e-mail: rimafuku@gifu-u.ac.jp

N. H. Hensel, P. Blessinger (eds.), *International Perspectives on Undergraduate Research*,
https://doi.org/10.1007/978-3-030-53559-9_6

versus honors students; and collaborative versus individual (Beckman and Hensel 2009).

This chapter describes the two main approaches to UR in Japanese higher education. The first approach is a product-centered model that has been developed traditionally in Japan. In Japan, research is generally embedded in the curriculum of undergraduate education. Final-year undergraduate programs in many institutions include a "research project and thesis writing" component, which is offered as a summative assessment of the undergraduate program or as a means of training young researchers in basic research practice. Generally, the focus is on research outcome and content, an important element of research-based teaching in Japan. This undergraduate educational trend can be traced back to the Meiji Restoration in 1868, when the feudal monarchy of the Tokugawa shogunate ended and control of the country returned to direct imperial rule under the Emperor. This historical event was a major turning point for all aspects of Japanese society, as it brought about modernization and Westernization of the economy, politics, military, and education. At that time, Japanese imperial universities were established based on certain Western models of higher education, including the Humboldtian concept that emphasizes the relationship between research and education (Altbach and Balán 2007). A product-centered view of research, therefore, has dominated undergraduate education for over a century.

The second approach to UR is a pedagogical approach focusing on student development and learning processes. In Japan, a movement toward valuing student-centered learning was triggered by social changes such as the expansion of higher education, declining birth rates, globalization, and the rise of the information society in the early 2000s. The revised version of the Basic Act on Education (2006), for instance, details the function of universities as follows:

> Universities, as the core of scholarship activities, shall cultivate advanced knowledge and specialized skills, inquire deeply into the truth and create new knowledge, while contributing to the development of society by broadly disseminating the results of their activities. (p. 5)

This concept of knowledge creation and inquiry processes as scholarly activity has much in common with the process of learning through research. In echoing the importance of students' active involvement in learning, in some disciplinary areas, including the fields of medicine and engineering, UR as student development is increasingly being offered to undergraduates. Moreover, in 2012, the Central Council for Education of the Japanese Government (2012) reemphasized the shift from an instructional paradigm to a learning paradigm in higher education. This has resulted in a sharp increase in the number of institutions seeking to develop and adopt active learning strategies, including problem-based learning, flipped classroom, and blended learning. Although UR as student development is yet to be widely achieved in Japanese higher education, some educators are gradually adopting UR as a new teaching strategy to promote a deeper approach to student learning.

The following sections will provide the details of these two approaches to UR in Japan, using examples of recent practice in undergraduate education. As the author currently engages in medical education, UR in medicine in Japan is central to this chapter. Additionally, UR in other fields including engineering and humanities is introduced to give an overview of UR in Japan. Given the current state of UR in Japan, the future development of UR will be discussed from the perspectives of management, curriculum development, teaching, and learning.

Undergraduate Research as Outcome Production

Final-Year Research Project and Graduate Thesis Writing

Owing to the aforementioned historical background of Japanese higher education, UR has been long viewed from a perspective of outcome production. For example, undertaking a final-year project and writing a graduation thesis have been widely regarded as the compilation of undergraduate learning. In fact, across all Japanese universities, 73.1% of humanities undergraduates and 87.1% of science and engineering undergraduates are required to write a graduation thesis (Kanda et al. 2013).

To complete a mandatory graduation thesis by the end of an undergraduate program, third (or fourth) year students across humanities and natural sciences are usually required to apply for and be assigned to a "seminar (*zemi*)" or a laboratory, which is a preparatory course for the research project they will undertake and the thesis they will write. A "seminar" functions as a learning community in which small groups of students, facilitated by a supervisor, discuss research content of interest, on an ongoing basis. This pedagogical method, based on the Humboldtian model, was first developed and adopted by Kyoto University School of Law in 1900 (Ushiogi 2008). Since then, "seminar" and the writing of a graduation thesis have been considered important elements of the integrated research activity of undergraduate education.

Previous research has highlighted several issues concerning "seminar" and graduation thesis writing within undergraduate programs. First, the faculty workload required for the supervision of student research projects is increasing. For example, a report that a faculty member in the humanities is required to supervise, on average, 16.6 students each year (Shinoda and Higeta 2013). Therefore, developing institutional support systems for faculty and staff is essential to ensuring the continued quality of UR. Second, the focus of assessment is conventionally greatly inclined toward the product of the research (e.g., what students found in their research and wrote in their thesis), rather than the process of student engagement with the research (e.g., how students engaged at stages of research planning, conducting a literature review, data collection, and analysis). The general criteria for assessing students' research-based learning include a logically structured argument, the significance of the study, and quality of the literature review in a product of graduation thesis (Wada 2014). In addition to the product of the research, the process of students' research activities needs to be assessed comprehensively to clarify the educational impact of "seminar" and graduation thesis.

Some pedagogical research has emphasized the importance of investigating students' "seminar" and graduation thesis writing learning experiences (Fushikida et al. 2011; Sakai et al. 2006; Yamada 2009). For instance, in a questionnaire survey conducted by Sakai et al. (2006), students perceived the learning outcomes of the final-year research project and graduation thesis as their understanding of the nature of research, synthesis of knowledge, active involvement in learning, and developing problem-solving skills. This process-centered perspective of student engagement with the graduation thesis could be the key to expanding the definition of UR in Japan.

UR for Researcher Training in Medicine

UR, as outcome production, also contains the dimension of researcher training. For example, developing a researcher training program in the fields of medicine and health sciences is a critical issue in Japanese higher education. To date, four Japanese researchers have been awarded Nobel Prizes in Physiology or Medicine and seven have received the Albert Lasker Award for Basic Medical Research. However, there are institutional concerns regarding whether leading researchers can continuously and successfully be trained in the areas of basic and clinical sciences. The reason for this concern is that many newly graduated doctors desire to work as clinicians and thus leave their universities or research institutions for positions elsewhere. As a result, the number of would-be physician-scientists working in universities is in decline (Shimizu 2011). Moreover, in the areas of basic and clinical medical sciences, the number of international journal papers produced by Japanese scholars has declined since 2008. Among G7 countries, Japan is alone in this trend, and Japanese scientific output is losing international competitiveness in the fields of basic and clinical medical sciences (Toyoda 2019).

In Japanese medical education, a six-year undergraduate program, in which students usually enrolled after completion of secondary education (see Imafuku et al. 2016), research-based teaching was originally integrated into the curriculum in the 1960s for those final-year medical students who wanted to become basic medical scientists or content experts. Until recently, the focus of research-based teaching was heavily focused on researcher training. In line with this trend, since 2002, medical schools at many research universities, such as the University of Tokyo and Tohoku University, have developed and offered MD-PhD credits as part of undergraduate programs. These programs allow undergraduate medical students to attend research courses to obtain postgraduate education credits, which function as a transition into PhD training. In other words, this is an opportunity for research-oriented medical students to earn both MD and PhD degrees in areas pertinent to medicine. However, since the number of applicants has been limited so far, the program is not yet regarded as a well-established core component of the medical education curriculum (Koibuchi 2018). Responding to this situation, in 2012, the Ministry of Education, Culture, Sports, Science and Technology (MEXT) decided to support leading programs to foster physician-scientists, as part of a project to promote undergraduate education. Many of the projects are extracurricular programs that do not require students to take a leave of absence from medical school. Students taking these courses receive research supervision in their third or fourth year of medical school along with attending postgraduate courses.

Furthermore, some undergraduate medical schools have independently developed research physician training programs. For instance, in 2007, Okayama University developed an Advanced Research Training Program, in which undergraduate medical students took postgraduate research subjects and participated in both residency and doctoral programs upon completing their undergraduate education (Okayama University School of Medicine 2019). Since 2015, Hyogo College of Medicine has offered a "research physician course," which allows medical students to concentrate on research activities in the third and fourth year of their undergraduate program by replacing lecture-based learning with self-study across all clinical medicine subjects (Hyogo College of Medicine 2019). Recently, as such programs have begun to receive government supports, researcher training methods have started to be developed. Therefore, the effectiveness of UR programs in training physician-scientists requires further examination.

Undergraduate Research as Student Development

UR in Engineering Education

In Japanese higher education, there is a growing belief of faculty members that besides developing research skills, participation in research activity has the potential to improve students' active/reflective learning skills and capacity for critical appraisal. In particular, early experience of research as an undergraduate is starting to be recognized by faculty members as an effective opportunity for student development. For example, UR as student development has been incorporated into the engineering curriculum in some Japanese universities. The Institute of Industrial Science at the University of Tokyo (2019) offers the Undergraduate Research Opportunity Program, which is an elective course for first- and second-year students. It emphasizes the learning of basic skills for and attitudes toward undergraduate research, through active involvement in frontline research as a member of a laboratory team. Students are given the list of possible research themes in laboratories, and can select a subject of their interest from a choice of 25 study areas on the list, including nanotechnology, information technology, biomedical sciences, and environmental sciences. Each year, six to ten students pursue this research course. Specifically, in this course, they are expected to autonomously develop research questions, research design, and data collection and analysis procedures in the selected area of study. Over the course of 12 weeks, each student is expected to complete an independent research project under the guidance of a supervisor. Their original research projects include: Auto-colorization of monochrome images using object recognition and Markov random field, long-term change in flow rate of a spring at a park in Tokyo and its possible cause, and rotational spectroscopy of HD by laser-induced fluorescence. At the end of the course, they are required to give an oral presentation or submit a final research report. Furthermore, opportunities of conference presentation and article writing are provided for students wishing to pursue their research after UROP.

Another example is a 12-week internship program to engage students in fieldwork research from the Faculty of Engineering at Gifu University. This program, developed in collaboration with 18 local manufacturing companies, encourages third-year students to investigate issues regarding the product development process, management, and systems in fieldwork. This inquiry-based course is underpinned by a process-oriented pedagogical approach to UR, which emphasizes the cultivation of transferable skills including communication, problem-solving, creative thinking, and critical appraisal. Each company accepts five to ten students. Past research projects have included developing a cooling and control system for casting material, a model for the analysis of motor oil flow, and creating assembly jigs using a 3D printer. At the end of this course, students are required to give a poster presentation about their research project. Students who completed the course have commented that they can now apply their knowledge to further research, problem-solving, and thinking about their future careers by relating to their research experience in the company (Gifu University 2017).

UR in Medical Education

Similarly, in the field of medical education, as the importance of facilitating the active learning of medical students is recognized, some medical schools have started to perceive research-based teaching as an educational opportunity for student development. Engaging students in research and inquiry is seen as helping cultivate the generic skills necessary for continuing professional development. Among Japan's 82 medical schools, 71 have implemented research-based courses in their six-year undergraduate programs (AJMC 2018). However, as there is insufficient research investigating student experiences and perceptions of UR, the educational impact of these courses on student learning has been not fully determined.

Gifu University School of Medicine, the author's own institution, has recent experience in educational practice and research on student experiences of UR. The School of Medicine provides a mandatory ten-week Research Experience course, in which all third-year students (approximately 110 each year) select a subject from 25 research themes from basic, social, or clinical medical sciences. Students are encouraged to actively pursue a research theme of interest in a scientific manner. It primarily involves project work, and there are no classes during the research weeks to distract from the inquiry-based student learning experience. Faculty members are expected to be involved in UR as facilitators of student learning through research. The details of course content and schedule, including research activities/tasks, formative assessment, and means of student–faculty interaction, can be designed at the discretion of faculty members and students in each research area. At the end of the final week, students are required to give poster and oral group presentations in front of all the third-year students and faculty members.

Research into medical education is among the 25 research experience themes. Each year, about five students select the theme of medical education. Class meetings of 2–3 h are typically scheduled three days a week, and the remaining class time each week (21 h) is allotted to self-directed research activity. In each class, students are expected to autonomously design a research and collect and analyze qualitative and/or quantitative data (see Imafuku et al. 2016). The research themes on medical education undertaken by the students in the past included medical students' views of work–life balance for physicians, medical students' inappropriate behavior in the classroom, medical student and faculty perspectives of the benefits of UR experience, and students' community medicine orientation and specialty choices. To date, students who conducted medical education research at Gifu University have given five presentations at regional conferences, and published four papers in regional journals, and published one paper in an international journal.

Investigating Medical Students' Experience and Perception of UR

Imafuku, Saiki, Kawakami, and Suzuki (2015) conducted a qualitative study to examine the manner in which medical students' perceptions of research and learning changed during their UR course in medical education. Data analysis of interviews with 14 students revealed that at the beginning of the course, the majority of students had a relatively narrow understanding of research, focusing on its content and outcomes. End-of-course reflections indicated increased attention to research processes including self-initiated learning activity, collaboration, and processes of knowledge construction. This research found that students' awareness of linking learning and research indicated an epistemological change, resulting in them adopting a deeper approach to UR learning. Students' revised perceptions of research included an understanding that it requires an inquiring mind, synthesis of knowledge, active participation, and collaborative and reflective learning.

Further research by Imafuku et al. (2018b) explored medical students' and faculty members' perceptions of the benefits of their UR experiences. Data analysis of interviews with 18 students and 11 faculty members showed that the faculty members commonly perceived transferable skills such as active learning and critical appraisal as UR benefits, while the students most commonly identified research-specific skills, such as conducting rigorous experiments and designing a study, and content knowledge as the most important benefits. Faculty members sought to enhance students' attitudes toward research during UR by encouraging inquisitiveness and being research-minded. This study revealed a gap caused by students' perceptions of research and faculty members' difficulties in engaging with students through research mentorship.

Students, therefore, perceived UR as an effective learning opportunity to enhance their research and transferable skills. In particular, epistemological changes in research and learning were found during the UR process (Imafuku et al. 2015). However, based on the findings that faculty members and students had different perceptions of the benefits of UR (Imafuku et al. 2018b), some practical issues regarding curriculum development need to be addressed. As there is limited Japanese research on the topic of the student UR learning process, further evidence for curriculum development is required.

Summary

Figure 6.1 gives a summary of the UR programs discussed in this chapter. Each program was described separately in the section of product-oriented or process-oriented view. However, it has to be noted that these views are not dichotomy but continuum. To the degree that the primary purpose of UR is to foster student learning, the emphasis might be on helping students to move along a developmental trajectory in the practice of research (Beckman and Hensel 2009, p. 40). In other words, the definition of UR that fits educational goals in an institution lies somewhere in the middle of this continuum. In the following section, I shall briefly discuss some issues for further development of UR in Japan.

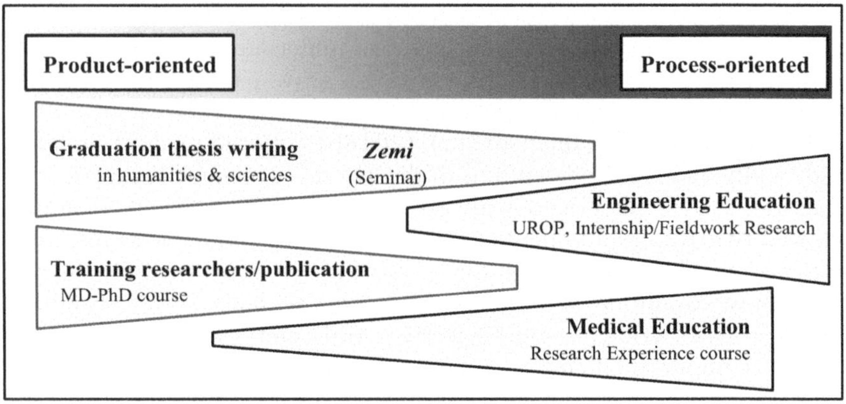

Fig. 6.1 A summary of the UR programs discussed in this chapter

Future Development

Historically, integration of research and learning has been an important issue in the enhancement of higher education in Japan, and to date, UR has generally been outcome-oriented, in the form of a final research project and graduation thesis. Recently however, engaging undergraduate students in research activities has been increasingly advocated by institutions as an innovative strategy to promote a high-impact learning experience. In other words, the process of research is emphasized as a pedagogical approach to undergraduate education.

Importance of Developing an Academic Community Supporting UR

Several issues must be resolved, including lack of a defining concept of UR, role ambiguity between students and faculty members, and necessity for effective assessment strategies for student learning. In order to do so, developing an academic community supporting UR among educators, staff, and students is essential. This is because currently, no organizations in Japan support UR, whereas the Council on Undergraduate Research was founded in 1978 in the United States. In this situation, each institution, department, or faculty member has to individually plan and develop UR policy, course design, and educational strategies. A limited number of universities in Japan provide research funding and travel grants for undergraduate students, but, as far as the author of this research has found, the government does not offer official financial resources for UR. Therefore, as Brew (2006) suggested, it is essential to the development of communities of practice to share the challenges of facilitating student learning, discuss the elements of UR best practice, and support programs among faculty and staff members in higher education institutions across Japan.

Developing an academic community would provide educators with an opportunity to define a concept of UR that fits the institutional mission. Educators must examine how they design and integrate student research activity within overall undergraduate education and how students learn in the designed curriculum over time. The constructive alignment proposed by Biggs and Tang (2011) can be an underlying concept when designing a holistic UR program. Educators must ensure coherence between intended learning outcomes, teaching, and assessment of UR within a four- or six-year undergraduate program.

Each institute should determine the roles of faculty members and students by aligning the faculty's intended learning outcome(s) with actual learning activities and student perceptions of the UR experience (Imafuku et al. 2018b). To achieve this, the UR course design must draw on a robust educational framework. For example, the Research Skill Development (RSD) framework proposed by Willison and O'Regan (2007) allows faculty members to define their own roles to suit evolving levels of student autonomy during each phase of research process. The research–teaching nexus framework developed by Healey and Jenkins (2009) shows how educators can shape student learning through UR. These frameworks offer tips for effective course design toward the achievement of the intended learning outcomes, including the content of research activity, balance between student engagement and faculty intervention, ways of giving reflective feedback, and summative assessment. As a first step to maximizing this opportunity, using Wenger's concept of communities of practice (1998), students and faculty members need to further discuss how they mutually interact and participate in an activity (i.e., mutual engagement), negotiate common purposes (i.e., joint enterprise), and share/develop UR resources (i.e., shared repertoire) during the UR program.

Inter-University Collaboration and Interdisciplinary Feature

Inter-university collaboration and interdisciplinary feature can be key strategies to enhance student learning in UR. As to inter-university collaboration in UR, for instance, a web research conference on medical education was held between Gifu University, Kagawa University, and Kansai Medical University during the Research Experience course. This conference stimulated student autonomous UR engagement. Imafuku, Nishiya, Saiki, and Okada (2018a) interviewed with students who participated in the conference, and one student reflected, *Discussing medical education research with students from other universities allowed me to critically evaluate my own project from different perspectives, which provided a future direction of my research.* Technology-mediated communication and/or face-to-face interaction among students working in the same research field promoted shared intellectual interests, knowledge sharing, and social relationship building in an academic environment (Imafuku et al. 2018a). As a next step, offering students the opportunity to conduct collaborative research with students from other universities is a pedagogical strategy that could enhance students' learning through research.

The interdisciplinary feature of UR can lead to diversity in student learning. As a trial, Gifu University provides an interdisciplinary UR course for first-year undergraduates. This course allows students from the Faculty of Education, Regional Studies, Medicine, Engineering, and Applied Biological Sciences to individually work on research projects beyond their major and to freely select a research theme of interest. For instance, an engineering student can research higher education policy and management, and a student from the regional studies can research Japanese literature. In the course registration, students are given a booklet that shows faculty members available for supervision and their research interests. Using this booklet, they need to find and contact a faculty member who matches the research field of their interest. Under supervision, students are expected to autonomously conduct their own research for four months and write up a research report by the final week. Among the submissions of final reports, the best research report award is selected by the University Academic Support Committee.

Conclusion

In Japan, UR has been conventionally implemented for a summative assessment of the undergraduate program or as a means of training young researchers (i.e., product-oriented view). As a relatively new trend, UR as student development is gradually adopted in Japanese higher education (i.e., process-oriented view). This is because UR started to be regarded as an educational strategy to promote a strong foundation for lifelong learning, including enhancing problem-solving skills, interpersonal communication skills, and intrinsic motivation toward learning. An academic community supporting UR, inter-university collaboration, and interdisciplinarity are key concepts for further development beyond the dichotomy of product- and process-oriented approaches.

Current Japanese society demands the ability to deal with complexity and uncertainty caused by globalization, population aging, information overloaded, and so forth. UR is an innovative way to engage students in meaningful learning to meet these societal demands.

Acknowledgment This work was partially supported by a JSPS KAKENHI Grant-in-Aid for Young Scientists (B) (no. 17 K15741). I would like to thank Dr. Mayumi Kawase (Faculty of Engineering, Gifu University) and Dr. Shio Kawagoe (Institute of Industrial Science, The University of Tokyo) for their help in understanding the recent practice of undergraduate research in engineering education in Japan.

References

Altbach, P. G., & Balán, J. (Eds.). (2007). *World class worldwide: Transforming research universities in Asia and Latin America*. Boltimore: Johns Hopkins University Press.

Association of Japanese Medical Colleges (AJMC). (2018). *Japanese medical education*. Tokyo: AJMC.

Beckman, M., & Hensel, N. (2009). Making explicit the implicit: Defining undergraduate research. *CUR Quarterly, 29*(4), 40–44.

Biggs, J., & Tang, C. (2011). *Teaching for quality learning at university–what the student does*. Berkshire: Open University Press.

Brew, A. (2006). *Research and teaching: Beyond the divide*. New York: Palgrave Macmillan.

Brew, A., & Boud, D. (1995). Teaching and research: Establishing the vital link with learning. *Higher Education, 29*(3), 261–273.

Central Council for Education. (2012). *Aratana mirai wo kizukutameno daigaku kyouiku no shitsuteki tenkan ni mukete* (Toward achieving a qualitative change in university education to build a new future). Tokyo: The Ministry of Education, Culture, Sports, Science and Technology. Retrieved from http://www.mext.go.jp/component/b_menu/shingi/toushin/__icsFiles/afieldfile/2012/10/04/1325048_1.pdf

Fushikida, W., Kitamura, S., & Yamauchi, Y. (2011). Gakubu san, yo nensei wo taisyou to shita zeminaaru ni okeru gakusyuusyayouin, gakusyuukankyou, gakusyuuseika no kankei (Analysis of relationships among learners' factor, learning environment, and learning outcomes in seminars for third- and fourth-year students). *Educational Technology Research, 35*(3), 157–168.

Gifu University Organization for Promotion of Higher Education and Student Support. (2017). *Learning support division newsletter* (Vol. 4). Retrieved from https://www.orphess.gifu-u.ac.jp/618522f3f9f6e657a9d573768d8d4d14.pdf

Healey, M., & Jenkins, A. (2009). *Developing undergraduate research and inquiry*. York: Higher Education Academy.

Hyogo College of Medicine. (2019). *Research course*. Retrieved from https://www.hyo-med.ac.jp/research_course.html

Imafuku, R., Saiki, T., Kawakami, C., & Suzuki, Y. (2015). How do students' perceptions of research and approaches to learning change in undergraduate research? *International Journal of Medical Education, 6*, 47–55.

Imafuku, R., Saiki, T., & Suzuki, Y. (2016). Developing undergraduate research in Japanese medical education. *CUR Quarterly, 37*(1), 34–40.

Imafuku, R., Nishiya, K., Saiki, T., & Okada, H. (2018a). Online and face-to-face: Developing an inter-university undergraduate research. *Medical Science Educator, 28*(1), 5–6.

Imafuku, R., Yasuda, S., Hashimoto, K., Matsunaga, D., Ohashi, Y., Yamamoto, K., Tsunekawa, K., & Saiki, T. (2018b). Exploring medical students' and faculty's perspectives on benefits of undergraduate research experience. *Medical Science Educator, 28*(3), 553–560.

Kanda, T., Tsuruma, K., Sato, M., Kojima, K., Shinoda, M., Higeta, T., et al. (Eds.). (2013). *Jinbunkei gakushikatei kyoiku ni okeru sotsugyou ronbun ga motarasu gakusyuuseika no kenshou* (Educational impact of working a graduation thesis in undergraduate education in humanities). Tokyo: Research Institute for Humanities, Gakushuin University.

Koibuchi, N. (2018). Medical research education. In Japan Society for Medical Education (Ed.), *White paper on medical education* (pp. 119–125). Tokyo: Shinohara Publishers.

Kuh, G. D. (2008). *High-impact educational practices: What they are, who has access to them, and why they matter*. Washington DC: Association of American Colleges and Universities.

MEXT. (2006). *Basic act on education*. Retrieved from http://www.mext.go.jp/en/policy/education/lawandplan/title01/detail01/1373798.htm

Okayama University School of Medicine. (2019). *ART program*. Retrieved from http://art-med.jp/

Sakai, H., Yamada, T., Shinto, T., Tanaka, K., & Araki, M. (2006). Kougakukyouiku ni okeru sotsugyou kenkyuu no yakudachi ni kansuru kouzou (Contribution of graduation thesis research toward engineering education). *Journal of JSEE, 54*(3), 51–56.

Shimizu, T. (2011). *Kisoigakukennkyusyabusoku no genjo to taisaku* (The shortage of researchers in basic medicine and the measures). Retrieved from http://www.mext.go.jp/b_menu/shingi/chousa/koutou/043/siryo/__icsFiles/afieldfile/2011/04/05/1303702_1.pdf

Shinoda, M., & Higeta, T. (2013). Jinbunkagakukei gakushikatei kyouiku ni okeru sotsugyouronbun no imizuke (The meaning of a graduation thesis in humanities departments). *The Journal of Management and Policy in Higher Education (Tokyo Univ.), 4*, 55–71.

The Institute of Industrial Science at the University of Tokyo. (2019). *Undergraduate research opportunity program*. Retrieved from http://www.oshimalab.iis.u-tokyo.ac.jp/UROP/

Toyoda, N. (2019). *Kagaku rikkoku no kiki. Shissokusuru nihon no kenkyuryoku* (A crisis of the science and technology-oriented country: A decline in Japanese capabilities to pursue scientific research). Tokyo: Toyo Keizai.

Ushiogi, M. (2008). *Funboruto rinen no shuuen? Gendai daigaku no shinjigen* (The end of Humboldtian model? New dimensions of modern universities). Tokyo: Toshindo.

Wada, M. (2014). Nihon no gakushi katei ni okeru kyouiku no ikkan toshiteno kenkyuu (Japanese undergraduate research and how to reform it). *The Journal of Learner-Centered Higher Education (Soka Univ.), 3*, 117–132.

Wenger, E. (1998). *Communities of practice: Learning, meaning, and identity*. Cambridge: Cambridge University Press.

Willison, J., & O'Regan, K. (2007). Commonly known, commonly not known, totally unknown: A framework for students becoming researchers. *Higher Education Research & Development, 26*(4), 393–409.

Yamada, Y. (2009). Zemikatsudou ni okeru Manabi wo saguru shiten to sono yuukousei (Examination to learning on the seminar activity: Seminar activity based on legitimate peripheral participation). *Kandai Psychological Review, 3*, 35–44.

CHAPTER 7

Undergraduate Research from Three Asian Countries

Enakshi Sengupta and Patrick Blessinger

Introduction

Undergraduate research (UR) research is the process of conducting original academic research on a specific topic by an undergraduate student. Although the institutionalization of UR is a relatively recent phenomenon, its roots can be traced back to the nineteenth century with the Humboldtian model of higher education (i.e., the integration of research and teaching) and the founding of the University of Berlin. The creation of MIT's Undergraduate Research Opportunities Program in 1969 is considered one of the first UR programs in the world. Undergraduate research provides opportunities for students to gain in-depth knowledge to learn and to conduct hands-on experiments and sharpen their problem-solving abilities.

E. Sengupta (✉) • P. Blessinger
International Higher Education Teaching and Learning Association,
New York, NY, USA
e-mail: ekapur@gmail.com; patrickblessinger@gmail.com

N. H. Hensel, P. Blessinger (eds.), *International Perspectives on Undergraduate Research*,
https://doi.org/10.1007/978-3-030-53559-9_7

Typically, during the undergraduate program, students are seldom exposed to academic research projects and wait until they reach their final year and then complete, for example, a capstone project or an internship program. As Evans (2010) pointed out, undergraduate students often lack the maturity and discipline to undertake and complete experimental work on their own. They need a certain amount of hand-holding by a research advisor—typically a faculty member or graduate research assistant who guides and mentors the student to complete the project. The mentoring process involves developing and understanding the research methodology, analyzing data, and effectively communicating their work in the form of written reports and presentations.

Wenderholm (2014) argues that it is difficult to assess the success rate of undergraduate research and to measure the positive impact on a student's life. According to Wenderholm (2014), the main challenge is to have motivated students with an adequate level of training who are interested in being involved in such projects. First-year students may have sufficient time, but their involvement also includes having a base level of knowledge. Thus, starting with second-year students could be more appropriate when involving them in undergraduate research.

Planning the project with measurable goals is another critical challenge. Wenderholm (2014) suggested that a second-year student could be involved in research so that they can learn from their mistakes and gain proficiency by the time they reach their final year. In their last term, students can have more time to be mentored by the faculty.

Boniak (2013) found that undergraduate research helps in generating self-motivation and self-directed learning and gaining new skills. Student research also helps enhance creativity among students. Faculty members, while dealing with students on a one-to-one basis, can dedicate more time to mentoring students. Mentors can vary their approaches and adopt research-informed, inquiry-based, and problem-based knowledge to apply it effectively. Mapolisa and Mafa (2012) also reported that undergraduate research eventually helps in improving the level of research in graduate students. They identified three levels of challenges faced by students, namely, mentor–student, student-related, and institution-related.

The mentor–student challenge involves engagement between the advisor and student, the time dedicated by the advisor and his/her availability, and student interest in the topic. Student challenges relate to personal issues in students' lives, such as time dedicated to research activities, financial matters, their level of motivation/commitment, and lack of knowledge. Lastly, institutional challenges involve overcoming hurdles like lack of research material, books, online resources that a student can tap into to enhance their experience. In most cases, students are willing to be involved in research if they receive incentives often based on credits offered for such projects, help in enhancing their GPA scores, or possibilities of financial reward or jobs.

The integration of undergraduates into the research process enables the faculty to break the monotony of classroom teaching and foster excitement in a project. Students are more inclined to become motivated learners as they realize that this kind of involvement will be beneficial to them in preparing them for their careers. While designing the study plan, the research component needs to be incorporated so that they do not clash with the existing course load or even the workload of the faculty members. The transition of being a mentee to the owner of the research project is a complicated process and may not have a required set of qualifying parameters. The research conducted by the student depends on the advisor's assessment of the student's capability to perform the tasks assigned to them successfully.

Literature Review

In 2005, the Council on Undergraduate Research, Washington, D. C., and the National Conference on Undergraduate Research jointly issued a statement recognizing undergraduate research as "the pedagogy for the twenty-first century", based on an inquiry-based model developed under a collaborative effort between mentee and mentor. Healey and Jenkins (2009, p. 3) argue: "All undergraduate students in all higher education institutions should experience learning through, and about, research and inquiry". Incorporating research into the curriculum exposes more students to such experiences (Jenkins and Healey 2012). Healey et al. (2014) proposed different roles that students undertake while being in a university, which was viewed as the concept of students as partners. Four such areas include: learning, teaching and assessment, pedagogic advice and consultancy, scholarship of teaching and learning, and subject-based research and inquiry.

The concept of students as researchers in a higher education context is an educational approach to support students in their engagement with undergraduate research to enhance their knowledge base and gain a more in-depth understanding of the subject. There are several definitions of undergraduate research, which are both informal and formal. A generally accepted definition, developed by the Council on Undergraduate Research, defines undergraduate research as "an active form of inquiry or investigation conducted by an undergraduate in collaboration with a faculty mentor that ultimately results in an original intellectual contribution to a larger body of knowledge" (Wenzel 1997). This definition is all-encompassing in nature. It includes research in both disciplinary and interdisciplinary fields, recognizing a teacher–scholar model that ensures that both students and faculty mentors have a mutual interest in the research experience.

The interpretation of undergraduate research tends to differ from discipline to discipline. Creative arts and design practice itself may constitute research (Yorke 2005), whereas the concept may vary in arts-based subjects with performances and exhibitions becoming a part of active research. In the sciences, the focus is based on experimentation and observation to generate data. Hence, Beckham and Hensel (2009) suggested that departments and institutions define undergraduate research based on the campus mission and disciplines.

Kuh (2008) recognized ten high-impact practices relating to undergraduate research based on the paradigm of students as researchers. The first-year course teaches students how to frame a research question and involves them in gathering data. Learning communities developed with students and faculty mentors can engage students in group research projects, and at a later stage, introducing writing-intensive courses embedded into the curriculum to help students practice writing of the research report. Collaborative assignments foster community practice among students. Working groups provide support to each other and enhance the research output. Service-learning opportunities and internships are various ways in which students can be inducted into research practices. These practices ultimately culminate in capstone projects and for some disciplines, may replace traditional program theses or reports.

Students-as-researchers is an active pedagogy that promotes students' appreciation for research as a discipline. The research process can be stimulated through assignments and giving students firsthand experience through live projects (Anderson and Priest 2014). Faculty members can develop higher-level thinking skills and critical judgment ability by examining how students gather and interpret data in light of their understanding. Faculty members encourage critical thinking skills and involvement in research, need to create time and space in the existing curriculum and assessment parameters that don't depend on rote learning methods but experiential, hands-on practices to encourage critical thinking skills and involvement in research, (Hodge 2011).

Awareness of students' willingness to participate in such research-based programs is essential for faculty members. Research Skill Development Framework of Willison and O'Regan (2013) provides a framework that contains six facets of research, enlisting the research process from initially clarifying a question, through to dissemination of the understanding generated by the inquiry. The research process links to the spiral curriculum (Bruner 1977), which states that the research process applies to each level of education, and the degree of rigor keeps elevating as students progress up the ladder of higher education.

Methodology

Faculty members of three countries, namely, India, Iraq, and Malaysia, where one of the authors has taught at the undergraduate level, participated in the study. Faculty members were chosen randomly from the author's past acquaintance. Twelve faculty members involved in teaching undergraduate students received a structured questionnaire. Ten of these faculty members were teaching undergraduates in the management department, and two of them belonged to the education department. Secondary data from documents available online belonging to the ministries of higher education in these countries added to the study. Answers were analyzed to form the narrative inquiry for the chapter. Through a thorough review of existing literature and qualitative studies, the benefits and impacts of undergraduate research on scholarly traits in students, as well as its effect on institutions, were studied.

Undergraduate Research in India

India's education system has been maintaining its status quo for over a few decades and with no real changes to the existing system (Altbach 2012). In most cases, the efforts have been fragmented and regional without creating an overall impact in significantly improving the current educational scenario. No institution is solely responsible for the lack of significant change. Students have graduated without being involved in a single research project, and as a result, these graduates lack the skills needed for their employability (Aspiring Minds 2018). With a population of 1.3 billion, there are only 216 researchers per million (UNESCO 2018), and research and development receive minimal support. In higher education, India's research expenditure is only 4% of GDP (UNESCO 2018). Estimates in 2018 suggest a relatively small number of the 161,412 students enrolled in doctoral research programs. Less than 0.5% of the total student enrollment in higher educational institutions, consisting of both public and private institutions, is pursuing graduate degrees (AISHE 2018).

Damini Saini of the University of Lucknow stated that "there is a considerate emphasis upon research for undergraduates in science area in India. All India Council for Technical Education (AICTE) is one of the Indian institutions which supports and funds the technical educational institutions for their conferences and research projects for life sciences, physical sciences, and chemical sciences, etc. The ministry of the government of India department of science and technology is providing fellowships for young researchers like start-up research grants for young scientists (e.g., early-career research award). Other than these, the Indian Space Research Organization (ISRO) also offers scholarships to young undergraduate scientists as projects and various training and development programmers at the undergraduate level like Bachelors of Engineering, Bachelors of Technology, and post-graduate levels like Masters in Engineering and Masters in Technology".

In March 2018, India launched the Prime Minister's Research Fellowship with an initial budget allocation of 16.5 billion Indian Rupees. Under the scheme, undergraduate and postgraduate students with a cumulative grade point average (CGPA) of at least 8.0 from elite Indian institutes such as the Indian Institute of Science (IISc), Indian Institutes of Technology (IITs), National Institutes of Technology (NITs), Indian Institutes of Science Education and Research (IISERs), and Indian Institutes of Information Technology (IIITs) will be eligible for direct admission in PhD programs of IITs and IISc.

Assistant Professor, Narendra Singh Chaudhary, Symbiosis Centre For Management Studies, echoed Saini's thoughts when he added that, "in India, there are funding agencies like the university grants commission and Ministry of Higher Education where a student can create a research proposal and get funding. The funding is specifically applicable to undergraduate studies. In every university or institution, internal funding supports undergraduate research but nothing specifically. Like, in my university, we get grants for minor research projects".

The Council on Undergraduate Research (CUR) defines undergraduate research (UR) as "an inquiry or investigation conducted by an undergraduate student that makes an original intellectual or creative contribution to the discipline". CUR is the apex body that oversees and affiliates institutes to perform undergraduate research on campuses. In an undergraduate study, students assist a faculty researcher, graduate student, and other undergraduates in researching areas of similar interests. Undergraduate students work as a support for collaborative research projects by either pursuing their research ideas or joining established (MIT 2018) research projects. Chaudhury clarified that, to his knowledge, no such definition had been coined in India to explain undergraduate research, "I don't feel there is any such consensus. The major focus of undergraduate studies is more on teaching, not research. Even the Ministry of Higher Education focus is the same. The undergraduate faculties are now asked to focus more on teaching than research".

Undergraduate research in India suffers to a great extent due to the system of affiliation (Sengupta 2019). As many as 500 colleges affiliated to a single university makes the entire system ungovernable (Altbach 2012). The system is a logistical nightmare leaving the colleges to function in isolation with no direct contact among each other. It defeats the very purpose of a university exchanging ideas, thoughts, and progress in the arena of research (Chandra 2007). Research is generally conducted in specialized institutes rather than in the colleges or universities in India (Sengupta 2019). Apart from working in isolation, there is little inter-disciplinary interaction when it comes to undergraduate research. Most colleges initiate undergraduate research only in the third year of study, reeling under the shortage of staff and unequal distribution of workload. Faculty workloads have not been adjusted to accommodate faculty time to supervise undergraduate research, nor undergraduate research valued in the promotion and tenure evaluation of faculty. Also, faculty and students do not coauthor papers at the undergraduate level.

Generally, in the case of research scholars, faculty write papers with them. "Our university doesn't have campus undergraduate research celebration days, and neither is the undergraduate research presented to political and government leaders. So generally, we do not have much support for the undergraduate students in social sciences in India in public universities", added Saini. While talking about faculty workload and undergraduate research, Chaudhury also added, "it is part of the job with no special accommodation or adjustments in the timetable. Faculties have to take out time from their normal working hours".

With a rich demographic dividend, India needs to chalk out a real concrete approach toward undergraduate research programs and initiate and motivate young learners to get actively involved in research studies at the very onset of their foundation course toward undergraduate studies. To tackle the dwindling number of researchers and the problems associated with low research output, it becomes imperative for both central and state governments in India to replicate a concept that has proven results in many other places across the world. The Indian education system has about 20 million first-generation learners, who will need systematic induction in research methods to use the vehicle of education as a tool to tackle real-world challenges. Moreover, the focus should remain on girls who need particular attention to encourage them to pursue fields in STEM (Science, Technology, Engineering, and Mathematics) (World Bank Brief 2015).

Undergraduate Research in Iraq

In the field of higher education, Iraq had played a vital role in the Middle East as one of the pioneers in this region before war ravaged the country's economy, along with its rich tradition in higher education. Data received from UNESCO (2004) postwar period showed that the country has more than 20 universities and over 50 technical institutes. With peace prevailing in the country, this number has grown and under the Ministry of Higher Education and Scientific Research (MHESR). The universities and several private colleges are offering courses in computer science, economics, and business management.

Iraq's first and largest university, Baghdad University, was founded in 1957, uniting several colleges that had been established earlier, including the College of Law (founded 1908), the Higher Teachers' Training College (1923), the College of Medicine (1927), the College of Pharmacy (1936), and the College of Engineering (1942). In the 1970s, the country showed a sudden growth of technical institutes, which were created to support the booming oil industry. These institutes were initially a part of the University of Baghdad but soon received their independent status. Every province was deemed to have their university to support higher education, which led to a further rise in the number of universities. The growth was necessitated to support equitable distribution of higher education in different geographical locations; however, the quality control and assessment of these universities were a cause of concern.

The Ministry of Higher Education plays a crucial role in determining the policy framework for research in universities. As early as 2003, a National Committee for Science and Technology, composed of university presidents, was formed that oversaw coordinating research activities. However, this committee did not have any significant impact in coordinating countrywide research activities, and such endeavors were shouldered individually by institutions. Industry partnerships and interactions with the economic sector of the country were handled personally by individuals based on their contacts (UNESCO 2004). "Unfortunately, neither the Ministry of Education nor the Ministry of Higher Education in the Kurdistan Region of Iraq has developed any policies to embed the undergraduate research into the curriculum. Ministry of Higher Education has taken a limited number of initiatives to incentivize research activities among scholars and graduate students. Still, more attention should be given to the undergraduate research since it would eventually build student capacity in developing critical thinking at an early age", as mentioned by Honar Issa, Secretary Board of Trustee, The American University of Kurdistan.

The thoughts of Karwan Sherwani, Head of Business and Management Department, Tishk International University, reflected the lack of any real coordinated effort. He commented that "No, our government is not very active in research at the undergraduate level and no reports or initiatives encourage innovation to support economic development available in record". Honar Issa further commented that "policies should be defined to make the process a necessity. There should be a strategic plan with set-up goals as well as a realistic timeframe to initiate the process of undergraduate research. In line with the policies that ministries develop, institutions should create a culture of research among undergraduate students by holding seminars and workshops that can address the significance of the process in the future career of students. Faculty members play effective roles in guiding and engaging students in research activities while highlighting needs and expectations". Fahrettin Sumer, Chair, Department of Business Administration American University of Iraq, Sulaimani, also stated, "I work for the American University of Iraq, Sulaimani (AUIS). Both the Kurdistan Regional Government (KRG) and the Iraqi central government have higher education ministries that regulate universities, but I have not heard of their support for undergraduate research".

Research centers had received generous financial support in the 1970s when the oil industry was booming, but the situation changed in the

1980s because of the conflict with Iran. The lack of funding, materials, equipment, and literature became even more evident in the 1990s when international sanctions prohibited the import of materials and equipment with possible dual-use. Given the budget shortages, research activities relied mainly on postgraduate students. They were often undertaken in cooperation with partners from the economic or military sector, which funded specific projects (UNESCO 2004). Mainstream education and primary research at the undergraduate level were not a priority of the war-torn country. Lack of funding and adequate resources restricted the presentation of papers.

The situation has now considerably improved. Universities are encouraging both undergraduate and postgraduate students to engage in active research projects, "In my institution, most classes require some research papers, but these are mostly based on library and online research. To a degree, they also engage the community by interviewing people outside the campus to prepare video projects and to do the assignments. Also, they are encouraged to do internships outside the campus", added Sumer. "Most of the universities and institutions advocate the undergraduate research, and some are implementing them and thinking very seriously of making it mandatory with regulation and credits. At the same time, some institutions are unable to implement very sound research projects due to the students' poor understanding, lack of infrastructure, and management support". He further added, "we have a student conference, and staff conference, in both we have students and lecturers co-authoring papers and presenting the paper in the international/national conference together".

University faculty members are encouraged to pursue research activities that might help in solving practical problems that beset society. Research had played a vital role in the Iraqi higher education system in the past. The effort needs to revive in a direction that supports the sustainable development of the country. Iraqis are now feeling the impact of globalization and the rapid growth of information and communication. Thus, research needs to be redefined in an Iraqi context supporting new and innovative approaches toward progress. The emphasis at the undergraduate level needs to be on faculty members as mentors and initiating young learners towards a research-oriented pedagogy instead of restricting the entire research agenda to a few scientific and dedicated centers.

Undergraduate Research in Malaysia

In the recent past, Malaysian higher education witnessed unprecedented growth. The last ten years saw a significant increase in the enrollment of students, an influx of international students, as well as growth in research publication, patents, and improving the standards and institutional quality. The drive and expertise of the Malaysian academic community government investments to stimulate research and innovation contributed to the growth. The Ministry of Higher Education realizes that the field of higher education is ever-growing, and further encouragement of robotics, internet, artificial intelligence, and disruptive technologies is needed. The Ministry is keen to prepare Malaysian youth to keep abreast of the latest developments and global trends, advocating research preparedness among students from the foundation and undergraduate levels. According to a faculty member who chooses to remain anonymous, "most institutions of higher learning in Malaysia offer undergraduate research as a compulsory module within undergraduate programs, and it is usually in the final year of the undergraduate program. In the university I teach at, XYZ University, it is usually taken by students in their final year of study. The module is called Final Year Project (FYP) and is completed over two semesters". In 2013, the Ministry of Higher Education in Malaysia prepared a Malaysia Education Blueprint 2015–2025 (Higher Education) or the MEB (HE), which showed research publication has increased threefold from 2007 to 2012; 70% of these articles have been contributed by the five Malaysian Research Universities (MRU) (MBE—HE 2015–2025). The number of patents filed also grew by 11% during this period, and universities have been successful in generating revenues through their research and consulting services. The anonymous faculty member added that "the Malaysia Plan framework charts the path for the nation's economic development. Innovation features as an important measure in the country's economic development. However, to my knowledge, there may not be a direct link between innovation to support economic development and undergraduate research".

Even though the country is excelling in its effort in promoting research and showcasing significant growth in the higher education sector, the ride to the path is not altogether smooth. Faculty members have often complained about the lack of enthusiasm, especially among undergraduates, lacking in critical thinking and communication skills. Students also lack in their ability to comprehend English or communicate in English, which hampers their ability to research in a field. Institutions that contributed data are trying to overcome such challenges and encourage students to come forward. "At my institution, we carry out research exhibitions and competitions to celebrate the innovations produced via undergraduate research. These exhibitions are sometimes open to the general public. We do share our innovation and research findings with government agencies, but this is primarily for research conducted by staff and postgraduate researchers".

Budgetary constraints and the rising cost of education have put a strain on many innovative projects and research work, especially encouraging research activities at an undergraduate level. Employers are looking at graduates with requisite knowledge, skills, and attitudes, and bridging the gap between market demand and the existing education system is becoming an arduous task. In Educational Inquiry (Extended Project), Module code: EDUC3028, offered by University of Nottingham, School of Education, Malaysia Campus (2018), the module has been "designed to enable students to learn critical thinking skills, where they will be taught some techniques for evaluating arguments rather than accepting what they read as the truth. Students can seek evidence, and by using their critical thinking skills, they would be able to judge the authenticity of the evidence. Learning how to do research should help sharpen their critical thinking skills. The module also should help in providing the skills to students to understand research articles written by eminent educators and other related relevant academic writings. This would enable them to gain knowledge from primary sources". Efforts are being made by universities to orient students toward a research-based curriculum from the very beginning of their undergraduate course so that they can move from a world of job seekers to a world of job creators (MBE—HE 2015–2025).

Recommendations

The emergence of the social sciences as a relatively new field of study, the refinement of educational research methods, and advances in brain research and learning science provide educators a broader set of tools and knowledge. Now they can determine the soundness of educational philosophies, teaching methods, and learning theories. Given the considerable potential for research of all types to enhance learning outcomes, higher education can benefit from integrating student research activities into curricula at all levels across all domains.

Research occurs along a broad spectrum, from scientific research to humanistic research to art-based research. This spectrum of research activities allows for considerable flexibility in integrating research into a variety of courses. By using both critical and creative thinking, students are better able to evaluate issues and problems from a more comprehensive set of perspectives. Principles and concepts from one domain or discipline are integrated with other domains or disciplines. All domains have the potential to inform and benefit other domains (Blessinger 2017).

Conclusion

As mentioned at the beginning of the chapter, the path to undergraduate research is not an easy task. Along with students' enthusiasm, time management, and faculty workload, there are issues of ownership, ethical practices, equality of opportunity, quality of supervision, and other resource constraints. The benefits of undergraduate research enhance the student–teacher relationship, the entire learning experience, and the culture of inclusion and openness. Undergraduate research should be encountered early in the student learning experience (Walkington et al. 2011).

Data revealed that little had been done to promote undergraduate research, mainly in the field of social science. Government funding in most countries is available to support research in science and engineering studies at the postgraduate level. In most cases, the curriculum does not include undergraduate research, and both students and faculty struggle to find a space for them. Faculty members teaching undergraduate students are overworked and often, no credits are given to them to accommodate undergraduate research or mentor students.

Undergraduate research should be embedded in the curriculum and can be scaled up to be inclusive. When universities adopt the pedagogical approach of student researchers, they assist students in preparing for their future careers. Students can inculcate a sense of ownership toward these projects. A research-based approach toward teaching will result in breaking the long-standing disconnect between teachers and students, providing rewards and recognition in support of the method.

References

AISHE. (2018). *All India survey on higher education 2017–18*. Ministry of Human Resource Development.

Altbach, P. (2012). The permanent crisis of Indian higher education. In P. Agarwal (Ed.), *A half-century of Indian higher education*. New Delhi: Sage Publications India.

Anderson, J., & Priest, C. (2014). Developing an inclusive definition, typological analysis and online resource for Live Projects. In H. Harriss & L. Widder (Eds.), *Architecture live projects. Pedagogy into practice*. Oxford: Routledge.

Aspiring Minds. (2018). *National employability report engineers – Annual report 2016*. Aspiring Minds, 2016.

Beckham, M., & Hensel, N. (2009). Making explicit the implicit: Defining undergraduate research. *Council for Undergraduate Research Quarterly, 29*(4), 40–44.

Blessinger, P (2017). Transforming learning through student research. *University World News*. Retrieved on 28th June 2019 https://www.universityworldnews.com/post.php?story=20170530085248669

Boniak, R. (2013). Overcoming challenges in field research for non-science majors. In F. Wilson & J. Thomas (Eds.), *Handbook for undergraduate research advisors* (pp. 123–129). Lanham: Rowman and Littlefield. 5 T.

Bruner, J. (1977). *The process of education* (2nd ed.). Cambridge: Harvard University Press.

Chandra, P. (2007). The decay of the campus. In *Building universities that matter* (p. 222). Hyderabad: Orient Blackswan.

Evans, D. (2010). The challenge of undergraduate research. *Association of American Colleges and Universities, Peer Review, 12*(2).

Healey, M., & Jenkins, A. (2009). *Developing undergraduate research and inquiry*. Higher Education Academy. Available at: http://www.heacademy.ac.uk/assets/documents/resources/publications/DevelopingUndergraduate_Final.pdf

Healey, M., Flint, A., & Harrington, K. (2014). *Engagement through partnership: Students as partners in learning and teaching in higher education*. York: HEA. http://www.heacademy.ac.uk/assets/York/documents/resources/publications/DevelopingUndergraduate_Final.pdf

Hodge, P. (2011). Revisiting 'how we learn' in academia: Practice-based learning exchanges in three Australian universities. *Studies in Higher Education, 36*(2), 167–183.

Jenkins, A., & Healey, M. (2012). Research-led or research-based undergraduate curricula. In D. Chalmers & L. Hunt (Eds.), *University teaching in focus: A learning centred approach* (pp. 128–144). Camberwell: Acer.

Kuh, G. D. (2008). *High-impact educational practices: What they are, who has access to them, and why they matter*. Washington, DC: AAC&U. 34 pp.

Malaysia Education Blueprint 2015–2025 (Higher Education). Ministry of Education, Malaysia.

Mapolisa, T., & Mafa, O. (2012). Challenges being experienced by undergraduate students conducting research in open and distance learning. *International Journal of Asian and Social Science, 2*(10), 1672–1684.

MIT News: MIT's much-imitated UROP turns 30. Massachusetts Institute of Technology, 2018., http://news.mit.edu/2000/urop-0202

Sengupta, A. (2019). *Improving research in India: Introducing undergraduate research in higher education.* ORF Issue Brief No. 277, February 2019, Observer Research Foundation.

UNESCO. (2018). *R&D data release.* UNESCO Institute for Statistics.

Walkington, H., Griffin, A. L., Keys-Mathews, L., Metoyer, S., Miller, W., Baker, R., & France, D. (2011). Embedding research-based learning early in the undergraduate geography curriculum. *Journal of Geography in Higher Education, 35*, 1–16.

Wenderholm, E. (2014). Challenges and the elements of success in undergraduate research. *Inroads, The SIGCSE Bulletin, 36*(4).

Wenzel, T. J. (1997). What is undergraduate research? *CUR Quarterly, 17*, 163.

Willison, J., & O'Regan, K. (2013). *Research skill development framework.* Retrieved 10 December 2013 http://www.adelaide.edu.au/rsd/framework/

World Bank Brief. (2015). *Educating India's children.* World Bank.

Yorke, M. (2005). *Issues in the assessment of practice-based professional learning.* http://surreyprofessionaltraining.pbworks.com/f/Assessment+of+Practice+Based+Learning+Mantz+Yorke.pdf

CHAPTER 8

Undergraduate Research in Brazil: A Study of the Contributions and Challenges for the Education of Young Researchers

Ana Lucia Manrique and Douglas da Silva Tinti

Introduction

Some studies investigate undergraduate courses, showing that the view on this academic space should be modernized. The reproduction and non-production of knowledge existing in this space are two of the problems the studies mention (Massi and Queiroz 2010). However, it is important to point out that Brazilian universities are anchored in the teaching-research-extension triad and in their inseparability (Maciel and Mazzilli 2010), and research is valued for providing diverse relationships between teacher, students and knowledge production. In this sense, it is understood that undergraduate research can make it possible to approximate and strengthen

A. L. Manrique (✉)
Mathematics, University of São Paulo, São Paulo, Sao Paulo, Brazil
e-mail: analuciamanrique@gmail.com

D. da Silva Tinti
Mathematics, Federal University of Ouro Preto, Ouro Preto, Brazil
e-mail: tinti@ufop.edu.br

N. H. Hensel, P. Blessinger (eds.), *International Perspectives on Undergraduate Research*,
https://doi.org/10.1007/978-3-030-53559-9_8

the relationship between teaching and research, and even more, between undergraduate and postgraduate studies (Pinho 2017).

This chapter presents an analysis of the investments made in the different regions of Brazil and in the various research areas made available by organizations that foster research in Brazil. It begins with a brief historical context of how undergraduate research consolidated over time, favouring the creation of different funding programmes. Therefore, this work explains the programmes that fund research developed by undergraduate students by providing them with a scholarship—financial support. Then, it explains the impact and the scope of the research works carried out in those programmes from 2001 to 2013. It closes with the presentation of an analysis of the investments made in the different areas of knowledge since 2001 by the Brazilian federal government.

Scientific Initiation in Undergraduate Courses in Brazil

Historically, funding of research undertaken by undergraduate students, called scientific initiation scholarship in Brazil, has been consolidated by the creation in 1951 of the National Council for Scientific and Technological Development (acronym CNPq in Portuguese). Although there had already been research activities with undergraduate students assisting professors since the 1940s (Bariani 1998), it was from the creation of the CNPq that annual scholarships began to be awarded to promote undergraduate research.

According to Bariani (1998), the Brazilian scientific initiation scholarship programme sought inspiration from experiences of the United States of America and France, which had activities of this type in an institutionalized way. Thus, in the 1950s and 1960s, the scientific initiation was installed in undergraduate courses in the various higher education institutions of Brazil. In the 1970s and 1980s, undergraduate research was strengthened and there was an incentive for the creation of graduate programmes. In 1988, CNPq created a programme exclusively dedicated to funding scientific initiation research works (in Portuguese, IC), called Institutional Scientific Initiation Scholarship Program (Portuguese acronym Pibic). This programme established a new way of awarding scholarships for scientific initiation research that were formerly granted directly to students, and not to the higher education institutions. Thus, the decade of 1990s was called the decade of scientific initiation in Brazil.

In this funding programme, scientific initiation grants were awarded directly to both public and private higher education institutions. Since then, HEIs have been responsible for selecting undergraduate students, by means of public edicts to be IC grantees and to conduct research guided by supervisors holding a PhD degree. The edicts are elaborated by committees organized inside the institutions and follow criteria established by the CNPq to select the research projects that are offered IC grants.

To be eligible for a Pibic grant, students must have excellent academic performance, and they must be advised by a full professor, who can, nevertheless, co-supervise in partnership with professors holding a master's degree. Therefore, the IC to be developed should be related to the supervisor's projects and lines of research.

CNPq also evaluates the Pibic in all institutions through annual events in which the scholarship holders present the results of their research, always counting on evaluators external to the institution. Currently, the CNPq's Institutional Scientific Initiation Scholarship Program (Pibic) has the following objectives:

(a) To awaken scientific vocation and encourage new talents among undergraduate students
(b) To contribute to reduce the average time for a candidate to obtain a master and a PhD title
(c) To contribute to the scientific training of human resources that will be dedicated to any professional activity
(d) To stimulate greater articulation between undergraduate and graduate studies
(e) To contribute to the training of human resources for research
(f) To contribute to reduce students' average time spent in graduate programmes
(g) To stimulate productive researchers to involve undergraduate students in scientific, technological and artistic-cultural activities
(h) To provide the grantee, advised by a qualified researcher, with the means to learn techniques and methods of research, as well as to stimulate the development of scientific thinking and creativity, arising from the conditions created by direct confrontation with research problems

(i) To expand students' access and integration into scientific culture (CGEE 2017, p. 6)

Thus, the Pibic implemented in Brazilian HEIs aims to: train future researchers and encourage the expansion and consolidation of research groups. Scientific initiation should also contribute to graduation courses, increasing the flow of undergraduate students to a master's degree, reducing the time of completion of master's and doctoral courses, and enhancing the quality of the works being defended.

In 2007, the CNPq instituted a new modality of scholarship for undergraduate research, the Institutional Program of Scientific Initiation and Technological Development Scholarship (Pibiti, in Portuguese). This program follows the same rules of Pibic, being offered to public and private HEIs. However, its specific objective is to encourage the students in activities of technological development and innovation processes.

The evaluations carried out by CNPq (CGEE 2017) show that Pibic and Pibiti are programmes to induce institutional policies on undergraduate research within HEI, besides strengthening the relationship between undergraduate and graduate courses. Also, over the years, the concept of scientific initiation has been constructed in higher education institutions as an activity carried out by undergraduate students under the guidance of a professor/supervisor, aiming at innovating the processes of teaching and learning, minimizing the distance between teaching and research. In addition, these activities intend to initiate the student in research practices and provide experiences in the development of the supervisor's research projects.

Currently, Pibic aims to provide undergraduate students with "learning techniques and research methods, while stimulating the development of scientific thinking and creativity, arising from the conditions created by direct confrontation with research problems" (CNPq 2006, para. 16). In this way, the students are expected to engage in scientific activities, promoting the development of scientific thinking, as well as generating and socializing the knowledge produced in their scientific initiation.

Furthermore, it should be considered that the supervisor has a relevant role in the process of scientific training of the students that hold the grants, mainly for the constitution of good researchers, familiar with the research process and with the scientific method, and for their qualification for graduate programmes (Bianchetti et al. 2012).

Besides CNPq, some states in the country have the financial support of other development agencies, called Foundations for Research Funding (FAPs). As examples, the São Paulo Research Foundation[1] (FAPESP) supports research carried out by students and researchers in the state of São Paulo, and the Minas Gerais Research Foundation[2] (FAPEMIG) plays the same role in the state of Minas Gerais.

Impact and Scope of Pibic in Brazil (2001 and 2013)

The evaluation report of Pibic in Brazil in 2017 (CGEE 2017) shows that the number of scholarships grew by 67% (from 14,500 to 24,300 scholarships) between 2001 and 2013. However, despite being a significant increase, this advance has not been accompanied by the increase in the number of enrolments in the country's undergraduate courses (which reveals a growth of 102% in the same period), indicating that there was no evolution in the number of Brazilian students enrolled in undergraduate courses with access to the Pibic (INEP 2002, 2015).

In fact, in 2001 there were 4.8 Pibic scholarships for every 1000 students enrolled, and in 2013 it decreased to 3.9 scholarships for every 1000 students. In other words, the growth in the number of IC grants did not follow the increase in the number of enrolments in higher education.

Another relevant aspect worthy of consideration is the regionality of scientific research. In the early years of the IC Programme, CNPq was specifically concerned with the disparities across regions. Brazil is divided into five regions, South, Southeast, Center West, Northeast and North, and one of the guidelines was that the Pibic should contribute to reduce regional differences in relation to the distribution of scientific competence in Brazilian territory. This recommendation was included in the guidelines of the Pibic until 2004 (CGEE 2017).

Figure 8.1 shows the percentage share of IC grants in the five Brazilian regions, in the years 2001 and 2013. It can be observed that the percentages of the North, Midwest and Northeast regions were the only ones that grew during the period presented. However, the percentages still show differences in relation to the regions.

For a more detailed analysis, information is needed on the number of enrolments in higher education during the study period. Figure 8.2 shows

[1] More information at: http://www.fapesp.br/

[2] More information at https://fapemig.br/pt/

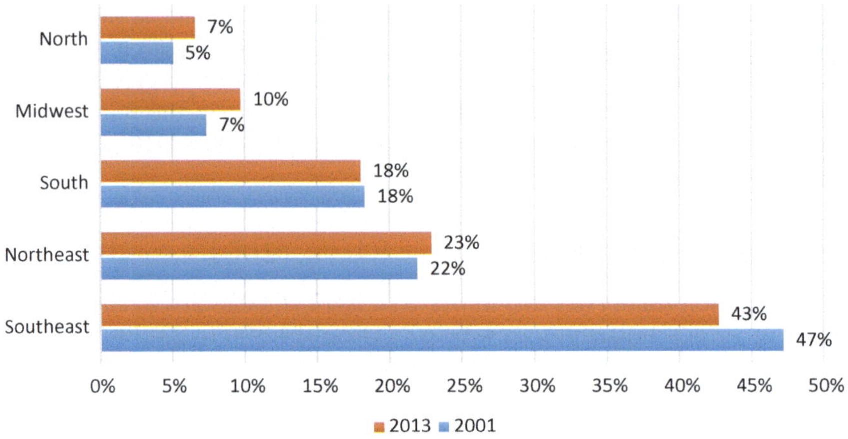

Fig. 8.1 Percentage share of the grants per year of the Pibic in the regions, in 2001 and in 2013. (Source: CGEE (2017))

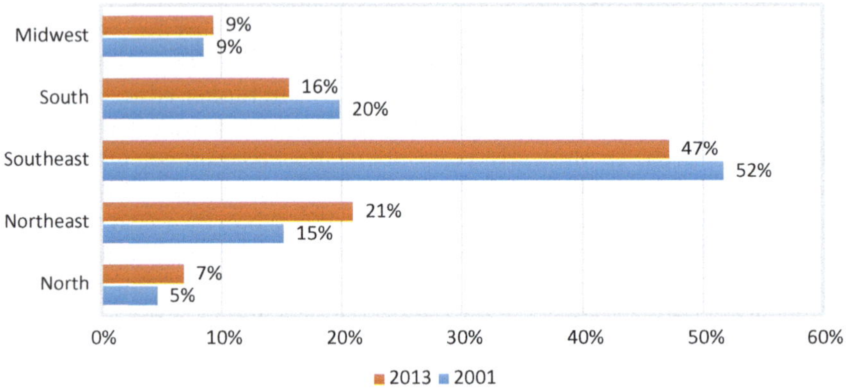

Fig. 8.2 Percentage participation of enrolments per year in higher education in the regions, in 2001 and in 2013. (Source: CGEE (2017))

the percentage share of enrolments in high education by region in the years 2001 and 2013. The figure shows that the North, Midwest and Northeast regions were also the only ones that presented an increased percentage in enrolments in higher education when compared to the number of IC scholarships, which can be observed in the Northeast region.

When comparing the enrolment numbers and the numbers of IC grants, a very large variation between the years 2001 and 2013 is evident. The distribution of Pibic scholarships computed by 1000 undergraduate students indicates that the differences remain in relation to the North and Northeast regions, as shown in Table 8.1.

The data in the table shows that although the number of IC scholarships for the North and Northeast regions increased, there was also a significant expansion of undergraduate enrolment in these regions, resulting in a decrease in the distribution of IC scholarships to the higher education students. In the Midwest region, variations in the number of enrolments and the number of Pibic scholarships were proportional, with the same number of Pibic scholarships remaining for every 1000 students enrolled from 2001 to 2013.

Regarding the gender of the scholarship holders, as shown in Fig. 8.3, in 2001, 55% of Pibic's scholarship holders were female, a figure that increased to 60% in 2013. That is, female participation in IC research projects has increased over the years.

The growth of female participation in undergraduate research projects is also evident in technological development projects and innovation processes such as the Pibiti. Figure 8.4 shows this growth in the period 2001–2013.

In 2007, the number of female students did not reach 40% of Pibiti's total scholarships, but by 2013, almost half of the number of technological initiation scholarships were awarded to female undergraduates.

Table 8.1 Distribution of Pibic scholarships computed by 1000 students enrolled by region, in 2001 and 2013

Region	*2001*	*2013*
North	5.2	3.8
Northeast	6.9	4.3
Southeast	4.3	3.5
South	4.4	4.5
Midwest	4.1	4.1

Source: CGEE (2017)

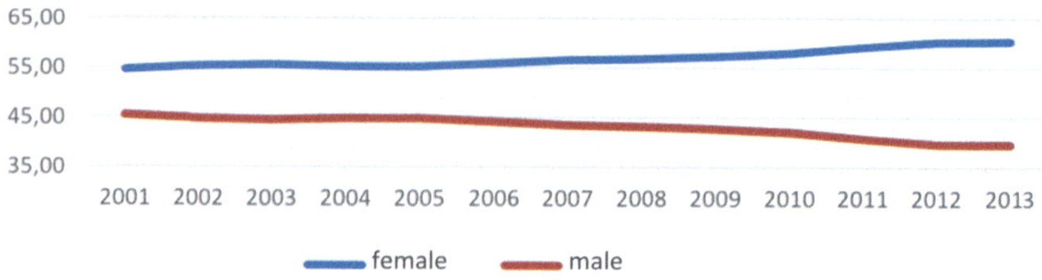

Fig. 8.3 Percentage distribution per year and sex of Pibic grantees between 2001 and 2013. (Source: CGEE (2017, p. 23))

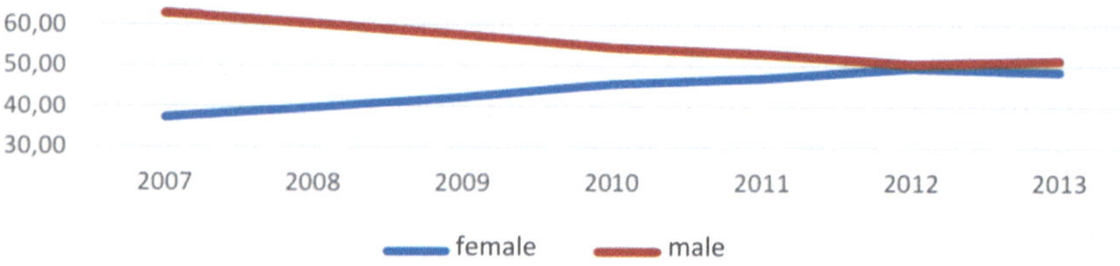

Fig. 8.4 Percentage share per year and sex of Pibiti grantees between 2001 and 2013. (Source: CGEE (2017, p. 23))

Current Statistics on Pibic in Brazil

Figure 8.5 shows the historical series of beneficiaries of CNPq's Pibic during the years 2001–2017. The data provided by CNPq indicate that from 2002 to 2014, the country registered a growth in the total number of beneficiaries of the scholarships awarded by the Pibic; the most significant growth being registered in the comparison between the total beneficiaries in 2008 (34,650) with the total beneficiaries in 2009 (38,413), which is 10.9% increase in the total number of beneficiaries.

As shown in Fig. 8.5, the most significant growth in investments represented by the number of Pibic beneficiary grantees occurred in 2006 and 2010. When comparing the investments in 2001 and 2017, there is an expressive growth of approximately 165% of the amounts invested. However, since 2014 the number of Pibic grantees has been decreasing in the funding scenario due to the distribution of resources to other programmes linked to CNPq and the economic crisis the country is still going through.

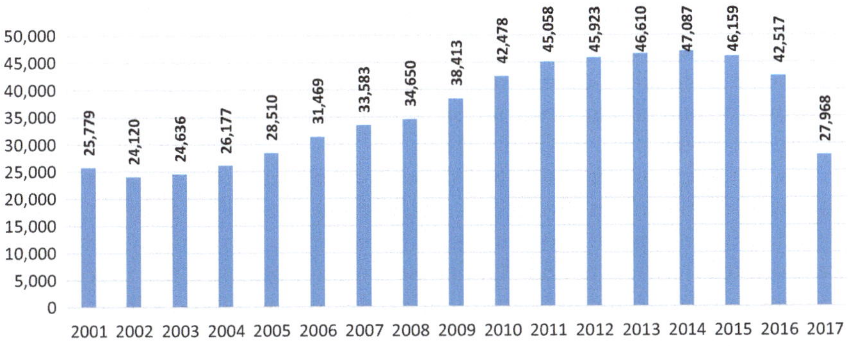

Fig. 8.5 Total beneficiaries of Pibic/CNPq in the period from 2001 to 2017. (Source: Adapted from the Panel of Institutional Programmes of Scientific and Technological Initiation (CNPq))

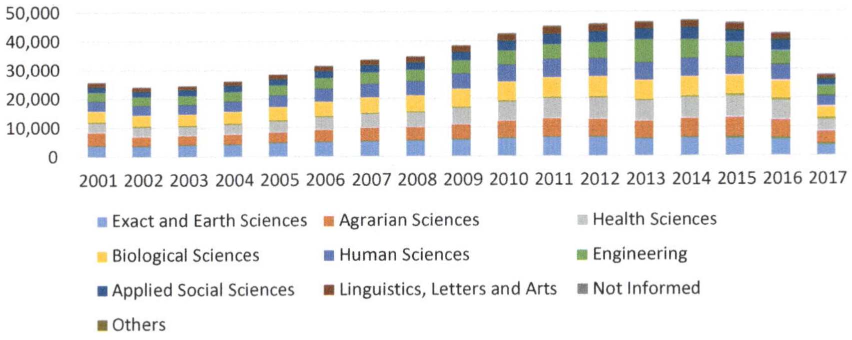

Fig. 8.6 Total beneficiaries of Pibic/CNPq, from 2001 to 2017, by area of knowledge. (Source: Adapted from the panel of Institutional Programmes for Scientific Initiation and Technological Development (CNPq))

Figure 8.6 presents the total number of scholarship holders from the CNPq's Scientific Initiation Programme during years 2001 through 2017, by areas of knowledge. The categories "Not informed" and "Other" are not depicted in Fig. 8.6 because the numbers are not representative.

The data in Fig. 8.6 reveal that all areas of knowledge were initially present over the years, with no significant decreases or increases in one are to the detriment of the others. Moreover, historically, the number of grantees grew in almost all areas during the period of increased investments for undergraduate research from 2003 to 2014.

Table 8.2 details CNPq's investments on Pibic by area of knowledge, showing the percentage distribution of investments by areas of knowledge during the years 2001 through 2017.

The data depicted in Table 8.2 shows a change in the percentage distribution of the resources invested in the different areas of knowledge. When considering the distribution of these investments, it is observed that the areas of Health Sciences and Applied Social Sciences show few percentage changes in relation to the investment carried out in CNPq's Pibic grants between 2001 and 2017. The Engineering area received an increasing amount of investment between the years 2011 and 2013, which was then reduced, and now it receives a percentage slightly above what it received in the year 2001. The area of Human Sciences received an increasingly smaller percentage over the years and the area of Exact and Earth Sciences received increasing percentages.

In 2001, the highest representation (18.45%) of the amounts invested in the Pibic was allocated to the Agrarian Sciences. However, in the other years, the highest figure is with the Exact and the Earth Sciences. Health Sciences and Engineering maintained their percentages over time, around 14% and 11% respectively, with few variations. Besides, there is an unfortunate decrease in representation in the area of Biological Sciences and Human Sciences. It is worth emphasizing that educational research studies are included in the area of Human Sciences.

Figure 8.7 presents a comparison of the representativeness of the Exact and Earth Sciences and Human Sciences areas from 2001 through 2017.

Figure 8.7 shows that the percentage of investments in the Human Sciences area has decreased, whereas the percentage of investments in the Exact and Earth Sciences increased substantially along the years.

When segmenting the total number of beneficiaries by gender, we observe that, historically, the number of females has always represented more than half the number of beneficiaries. This relationship can be seen in Fig. 8.8.

Figure 8.8, also shows a significant increase in the number of female IC grantees over the years, mainly during 2008 and 2016, followed by a reduction revealing that in 2017 the number of beneficiaries had returned virtually to the figures of 2001.

It was highlighted at the beginning of the chapter that besides the undergraduate research grants made available by CNPq's Pibic, most

Table 8.2 Percentage distribution of investments per year and by areas of knowledge

Areas										
Year	*Exact and earth sciences*	*Agrarian sciences*	*Health sciences*	*Biological sciences*	*Human sciences*	*Engineering*	*Applied social sciences*	*Linguistics, letters and arts*	*Not informed*	*Other*
2001	16.13	18.45	13.05	15.57	14.06	10.66	6.97	4.72	0.35	0.04
2002	17.03	13.56	14.04	16.80	14.21	11.95	7.58	4.68	0.03	0.12
2003	17.20	13.65	13.76	16.72	14.10	11.74	7.89	4.73	0.00	0.21
2004	17.44	13.61	13.51	16.58	14.32	11.72	7.78	4.71	0.00	0.33
2005	17.55	13.26	13.69	16.81	14.54	11.52	7.62	4.69	0.00	0.32
2006	17.07	13.35	14.35	16.93	14.33	11.50	7.42	4.73	0.00	0.32
2007	16.84	13.61	14.52	16.61	14.40	11.75	7.33	4.61	0.00	0.33
2008	16.80	13.37	14.64	16.67	14.48	11.69	7.31	4.62	0.00	0.42
2009	17.93	13.18	14.73	16.08	14.33	11.51	7.19	4.64	0.00	0.41
2010	20.89	13.64	14.24	14.55	13.29	11.61	6.77	4.43	0.06	0.52
2011	19.54	14.25	14.14	14.38	13.06	12.19	7.13	4.55	0.01	0.75
2012	19.61	14.00	14.10	14.36	12.95	12.65	7.24	4.22	0.02	0,85
2013	19.40	13.39	13.38	13.63	12.04	16.84	7.02	3.68	0.01	0.61
2014	20.13	14.11	13.74	13.48	12.11	14.43	7.39	3.97	0.00	0.64
2015	20.79	14.77	14.50	13.50	12.34	12.06	7.49	3.98	0.00	0.57
2016	21.14	15.00	14.73	13.82	11.91	11.61	7.43	3.95	0.00	0.41
2017	20.71	15.14	14.82	13.70	12.15	11.98	7.38	4.08	0.00	0.04

Source: Adapted from the Panel of Institutional Programmes of Scientific Initiation and Technological Development (CNPq)

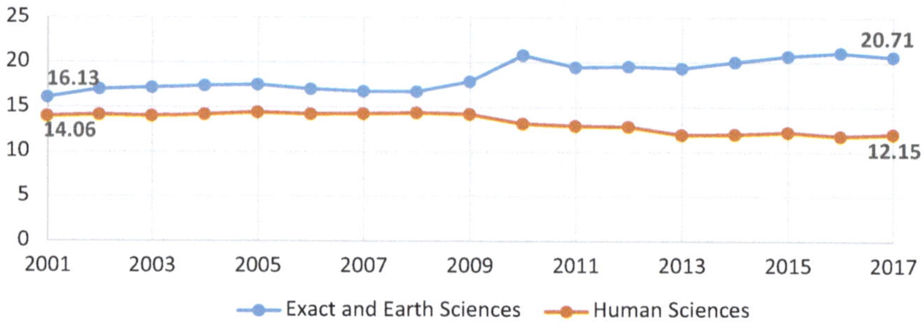

Fig. 8.7 Percentage distribution of investments in the areas of Exact and Earth Sciences and Human Sciences from 2001 through 2017. (Source: Adapted from the Panel of Institutional Programmes of Scientific Initiation and Technological Development (CNPq))

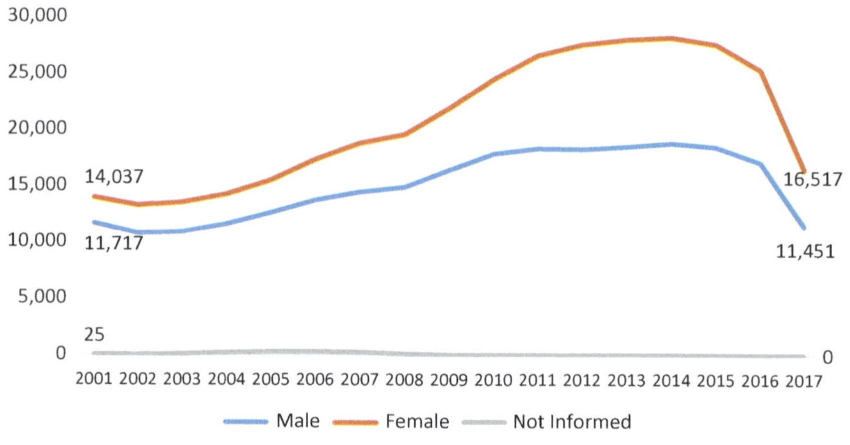

Fig. 8.8 Segmentation by sex of total beneficiaries of Pibic/CNPq from 2001 through 2017. (Source: Adapted from the Panel of Institutional Programmes of Scientific Initiation and Technological Development (CNPq))

Brazilian states also have research funding agencies, the Foundations of Research Funding (FAPs). To estimate the contribution of the FAPs for the development of undergraduate research, we collected data from both FAPEMIG, from the state of Minas Gerais, through a report made available by the agency itself (FAPEMIG 2018); and FAPESP, from the state of São Paulo, through data presented in their web page.

In 2018, FAPEMIG awarded 3999 scholarships to undergraduate students linked to Pibic. FAPESP awarded 1101 scholarships for scientific initiation. Hence the importance of the FAPs in the Brazilian context for the strengthening and financing of scientific initiation research in undergraduate studies.

Final Considerations

This chapter has presented and discussed information on the investment made by the Brazilian government in research in undergraduate courses, called scientific initiation research programme (IC). Financial support is granted annually to the best students, who are selected by research projects, guided by supervisors holding a PhD degree, from the different areas of knowledge.

Some studies have also investigated the contributions of scientific initiation to the students' academic development. Massi and Queiroz (2010), in a review of research about IC, found that some authors identify scientific research in motivating students in the process of acquiring knowledge, helping them to have better performance in the several disciplines of the course. It can also enable students to acquire scientific and specific knowledge in their undergraduate course, favouring a comprehensive training and reducing the detachment from the curriculum structure of their undergraduate course. Typically, undergraduate students complain of excessive content, often transmitted in lectures that are meaningless and unrelated to the undergraduate course.

Massi and Queiroz (2010) also highlight some of the skills developed by the scholarship holders involved in research projects, which were pointed out by the studies on the IC, namely, critical reasoning, autonomy, creativity, responsibility, analytical and interpretative capacity, leadership, facility in interpersonal relationships, altruistic values, self-worth, and self-esteem. They also emphasize that the development of IC research activities guided by PhD professors in a research group makes it possible for the grantees to understand the trajectory of academic career and this may, as a consequence, influence their professional choice.

Moura (2018) indicates that the IC experience is unique, because, while few Brazilians access higher education and scientific capital, given the historical selectivity and inequality of education, the research experience makes students establish a *habitus* that will allow for some accumulation of capital that is necessary for them to remain in, compete in and reproduce the university field.

Bariani (1998) also points out how important it is that all undergraduate students develop scientific skills, whatever their future profession, not just for an academic career. Scientific skills are associated with a process of teaching and learning that prioritizes the production of knowledge, that is, an education that is not based on simple presentation of information.

However, considering Brazil's current political and financial crisis, the country's main agencies aimed at developing and funding scientific initiation research have been drastically affected. Some of them have suspended notices and grants that would fund this research from 2019 through 2020.

This context has also had an impact on the funding of all scientific research throughout the country. Nevertheless, given the importance that scientific research represents for the scientific and technological development of a nation, we defend this public policy focused on the development of scientific research, aiming at its preservation, continuity, evaluation and availability of material and funding resources.

References

Bariani, I. C. D. (1998). *Cognitive styles of college students and scientific initiation*. Doctoral thesis, Universidade Estadual de Campinas, Campinas.

Bianchetti, L., Silva, E. L., Oliveira, A., & Turnes, L. (2012). Initiation to research in Brazil: Policies for educating young researchers. *Educação (UFSM), 37*(3), 569–584.

CGEE – Centro de Gestão e Estudos Estratégicos. (2017). *A Formação de novos quadros para Ciências, Tecnologia & Inovação: avaliação do programa institucional de bolsas de iniciação científica (Pibic)*. Brasília: Brasil Government Printing Office.

CNPq – Conselho Nacional de Desenvolvimento Científico e Tecnológico. RN017/2006. (2006). *Programa Institucional de Bolsas de Iniciação Cientifica – PIBIC*. Retrieved from http://www.cnpq.br/view/-/journal_content/56_INSTANCE_0oED/10157/100352#rn17062

CNPq – Conselho Nacional de Desenvolvimento Científico e Tecnológico. *Painel dos Programas Institucionais de Iniciação Científica e Tecnológica*. Retrieved from http://www.cnpq.br/web/guest/painel-programas-institucionais-de-ict/.

FAPEMIG – Fundação de Amparo à Pesquisa do Estado de Minas Gerais. (2018). *Relatório de atividades de 2018*. Retrieved from https://fapemig.br/media/media/Relatorio_Fapemig_2018.pdf

FAPESP – Fundação de Amparo à Pesquisa do Estado de São Paulo. *Biblioteca Virtual da FAPESP*. Retrieved from https://bv.fapesp.br/pt/

INEP – Instituto Nacional de Estudos e Pesquisas Educacionais Anísio Teixeira. (2002). *Censo da Educação Superior 2002: Resumo Técnico*. Brasília: Brasil Government Printing Office.

INEP – Instituto Nacional de Estudos e Pesquisas Educacionais Anísio Teixeira. (2015). *Censo da Educação Superior 2013: Resumo Técnico*. Brasília: Brasil Government Printing Office.

Maciel, A. S., & Mazzilli, S. (2010). *Indissociabilidade entre ensino, pesquisa e extensão: percursos de um princípio constitucional*. Proceedings from Reunião Anual da ANPEd, 33, 2010, Caxambu, MG. Retrieved from http://33reuniao.anped.org.br/tp33.htm#gt11

Massi, L., & Queiroz, S. L. (2010). Studies on undergraduete research in Brazil: A review. *Cadernos de Pesquisa, 40*(139), 173–197.

Moura, D. D. N. (2018). *Contribuições do PIBIC/CNPq para a constituição do habitus de pesquisador*. Master's dissertation. Universidade Comunitária da Região de Chapecó, Chapecó.

Pinho, M. J. (2017). Science and education: Contributions of the scientific initiation in higher education. *Avaliação* (Campinas) [online], *22*(3), 658–675.

CHAPTER 9

Finding Inspiration: Sharing Practice and Developing Authentic Multimedia Artifacts for Supervision of Undergraduate Research in Irish Higher Education

Roisin Donnelly, Claire McAvinia, and Claire McDonnell

Introduction

Good supervision is essential in ensuring successful outcomes for undergraduate research students, yet few new supervisors receive training for this role (Roberts and Seaman 2017; Healey and Jenkins 2018). In the

R. Donnelly (✉)
Business, TU Dublin, Dublin, Ireland
e-mail: roisin.donnelly@TUDublin.ie

C. McAvinia
Learning, Teaching and Technology Centre,
TU Dublin, Dublin, Ireland
e-mail: claire.mcavinia@tudublin.ie

C. McDonnell
School of Chemical and Pharmaceutical Sciences,
TU Dublin, Dublin, Ireland
e-mail: claire.mcdonnell@tudublin.ie

N. H. Hensel, P. Blessinger (eds.), *International Perspectives on Undergraduate Research*,
https://doi.org/10.1007/978-3-030-53559-9_9

context of Irish higher education, there have been calls for investment by higher education institutions (HEIs) into provision of suitable supervisor professional development opportunities. Since the late 1990s, the research landscape in Ireland has developed very significantly, underpinned by the recognition that talented people are at the heart of any national innovation system. Significant investment has resulted in Ireland ascending in international rankings of research capacity (HEA 2017). This is a positive development in postgraduate supervision. However, the majority of faculty in Irish HEIs have undergraduate supervision roles without the professional development (PD) opportunities that exist to support masters and doctorate level supervision. Hanratty, Higgs and Tan (2011, p. 37) have observed that 'academic staff who are attempting to initiate change in undergraduate teaching and learning strategies are often working in isolation within centers where postgraduate disciplinary research dominates the agenda'. Rowley and Slack (2004) have argued for a proactive approach to supervisor development. This study reports on a module that has been developed as such a proactive form of PD for undergraduate supervisors in Irish higher education.

The authors are faculty developers and academics in a new Technological University in Ireland, TU Dublin. In common with others (Roberts and Seaman 2018), we have found plentiful research into the supervision of PhD students and some on masters projects, but much less to draw on for supervisors at the undergraduate level. Over the last five years, there have been a number of publications in the area of undergraduate research mentorship. The term applied to the student–faculty member relationship in Ireland, and also in the UK and Australia, is 'supervision' and mentoring can extend beyond a professional supervision relationship to a personal one (Larson et al. 2018). Even so, these recent publications have been a welcome addition that can be drawn from. In Ireland, as is the case in the UK, all honors degrees usually incorporate a capstone research project. The module at the heart of this study was designed to support both novice and more experienced undergraduate supervisors. It has been running since 2015 and is entitled 'Supervising Undergraduate Dissertations and Projects'.[1] The module forms part of an accredited postgraduate program for faculty—the MSc in Education offered in TU Dublin.

[1] http://www.dit.ie/aadlt/lttc/academicdevelopment/postgradcpd/supervisingugdissertationsprojects/

In this chapter, we present the specific details of the module and the national context in which it is situated. We then present an evaluative study of the impact of the module, exploring how sharing supervision practice as well as the production of a multimedia resource have helped supervisors of undergraduate research find inspiration from each other. We discuss participants' perceptions of what makes excellent undergraduate research, the implications of this for their practice, and the place of the module in shaping that practice. We conclude by identifying the benefits of the module and aspects of supervisors' PD that can be further developed and supported.

National Context and Rationale for PD in Undergraduate Supervision

Nationally and internationally, undergraduate research has become more prominent in recent years. In Ireland, research is an important element of most undergraduate degree programs across the disciplines. The National Framework of Qualifications (2003) sets guidelines for the definition of undergraduate program learning outcomes at the honors degree level that point toward the inclusion of undergraduate research projects and dissertations. These requirements include (NQAI 2003, p. 17):

- Detailed knowledge and understanding in one or more specialized areas, some of it at the current boundaries of the field(s).
- Demonstrate mastery of a complex and specialized area of skills and tools; use and modify advanced skills and tools to conduct closely guided research, professional or advanced technical activity.
- Use advanced skills to conduct research, or advanced technical or professional activity, accepting accountability for all related decision making; transfer and apply diagnostic and creative skills in a range of contexts.
- Act effectively under guidance in a peer relationship with qualified practitioners; lead multiple, complex and heterogeneous groups.

The Irish National Framework of Qualifications has been approved as being compatible with the Qualifications Framework for the European Higher Education Area (Quality and Qualifications Ireland 2006), which means that it is consistent with the European bachelor, master and doctorate cycles (Bologna Working Group 2005).

Undergraduate research is disseminated locally in institutions through events and exhibitions, but also nationally through conferences and seminars. Examples are the Science Undergraduate Research Conference and the All Ireland Conference of Undergraduate Research.[2] Similar conferences and events take place in the UK, and have been observed with interest. Faculty active in such events in the UK have visited TU Dublin to speak about their work, including contributing inputs to our module on topics that have broad resonance, such as strategies for good practice throughout the supervision lifecycle, and linking teaching and research throughout the curriculum.

The module we developed is the first of its kind in the Irish HE sector. Since its inception, there have been participants from across the HE sector in Ireland, although the majority have been based in Dublin or the surrounding areas. As such, it can be regarded as representative of the issues facing UG supervisors nationally. This national audience includes supervisors new to their role, from across the disciplines, as well as more experienced supervisors who wish to share and expand their knowledge and experience in the UG research domain. Participants joining together in their learning in this module from institutions across the sector means that a sharing of different institutional regulations and practices, roles, expectations and responsibilities of the UG supervisor takes place. It is especially insightful for participants from many different contexts to have a space where they can collectively acknowledge the challenges of UG supervision and the accompanying assessment process. The module supports this range of UG supervisors from across the disciplines and institutions to reflect on their supervision practice for both pedagogic and professional development reasons, whilst cultivating scholarly exchange by encouraging them to share and critique dialogues about UG supervision.

Colleagues in two other HEIs have recently validated modules relating to undergraduate supervision and there are plans to commence these shortly. Also, a Digital Badge for supervision of postgraduate research has recently been developed by the National Forum for the Enhancement of Teaching and Learning, aligned with Ireland's National Framework for Professional Development (2016), and this may be used by faculty with

[2] http://sure-network.ie/conference/
https://www.ul.ie/ctl/events/all-ireland-conference-undergraduate-research-aicur

wider-ranging supervision responsibilities. Provision in some institutions allows for supervisors of undergraduate research to join modules or workshops intended for supervisors of postgraduate research, but this approach does not seem to be widespread. However, the argument for tailored development activities for supervisors of undergraduate research emerges clearly from the literature, as we explore further in the next section.

Demand for Professional Development for Undergraduate Supervisors

A review of the literature in faculty development and undergraduate research highlights a series of calls for professional development around a number of themes. Educational literature acknowledges the value for academics inquiring into and critically reflecting on their professional practice. Wisker (2012) has drawn attention to the need for supervisor professional development in the light of diversity, change and demand from one subject, one institution and one supervisor to another. She argues that both supervisors and institutions need to focus on supervisory developmental needs and practices. We are in agreement that the role is now more visible, and needs clarification and development for faculty, recognizing differences from one discipline to another and one supervisor to another. As Wisker points out, many faculty members perform this role but there are few opportunities to reflect on, develop or share good practice with others. This line of thinking informed the approach we undertook to the development of the PD module.

There have been more recent calls for supervisors who have been trained in mentorship (The Guardian 2017; Moore and Felten 2018). Additionally, supervisors of undergraduate research face challenges in a context where the ethos of support and well-being in relation to students is arguably at an all-time high (Wynaden et al. 2013). We sought to recognize the potential value of peer support to build confidence, as this has previously been identified as important in academic development (Boud 1999; Warhurst 2006) and building professional confidence.

There have been extensive guidelines produced to support supervisors and students in research, particularly at postgraduate level, nationally and internationally (Lee 2012; Wisker 2012). A new 'Supervising Postgraduate Research', SEDA course is on offer in UK institutions for new PG supervisors, which runs twice a year over two intensive days, supported by online activities and a portfolio. In addition, the Research Supervision Recognition Programme is a professional development toolkit that includes the sector-approved 'Good Supervisory Practice' framework and offers a route to recognition specifically for research supervision, from the UK Council for Graduate Education (2019). Comprehensive work by the National Academy for Integration of Research, Teaching and Learning (NAIRTL 2012) in Ireland had a particular focus on developing a framework to provide training and support for academic supervisors of research postgraduate students, including workshops, short courses and other initiatives. Although NAIRTL is no longer active, it previously worked with Irish higher education institutions to develop, implement and advance effective research-informed teaching and learning practices to enhance the student learning experience at undergraduate (Hanratty et al. 2011) and postgraduate levels. To this end, NAIRTL has initiated a wide range of events and activities that support stronger links between research and teaching (NAIRTL 2011).

In Vereijken's (2017) study on novice supervisors' practices, analysis revealed four kinds of dilemmas that may influence research supervision practices, namely, questions regarding regulation, student needs in relation to supervision, the student–supervisor relationship and supervisors' professional identity. Further afield, the scholars' conversations in Larson, Partridge, Walkington, Wuetherick and Moore (2018) of key terms, concepts and initiatives in mentored undergraduate research and inquiry in different international contexts were helpful in shaping our own local practice.

In terms of the topics that PD needs to explore for supervisors, a focus on the supervisor–supervisee relationship remains paramount. This relationship can be awkward and confusing, and sometimes uncomfortable and challenging (Grant et al. 2012). In Irish HE, there is not a formal body that guides staff in best practice in undergraduate supervision, and many programs that do exist for faculty are optional. The professional relationship between supervisor and student has received significant consideration in the literature. Rowley (2000) argues that the underlying philosophy is that supervision is a partnership between student and supervisor. Wisker (2012) encourages supervisors to reflect on and enhance their research supervision practice with a diversity of students on a variety of research projects. The student–supervisor relationship and style of supervision has also been previously investigated at undergraduate level with Hammick and Acker (1998) in particular exploring knowledge flow and power dynamics. As part of this valuable relationship, feedback has been identified as playing an important role. A study by Baker, Cluett, Ireland, Reading and Rourke (2014) reported that 88% of students reported peer supervision to be helpful, with themes being 'support and sharing', and 'progress and moving forward'.

Of particular relevance to our PD module, disciplinary perspectives in supervision have also been the source of research in previous years. Zydney, Bennett, Shahid, and Bauer (2002) analyzed the perceptions of 155 science and engineering faculty in a university with an extensive undergraduate research program. Faculty thought the undergraduate research experience provided important educational benefits to the students, in agreement with results from an alumni survey. Faculty who supervised undergraduates for a longer period of time and who modified their research program to accommodate undergraduates perceived a greater enhancement of important cognitive and personal skills. Within the discipline of social science, Todd, Smith and Bannister (2006) investigated the experiences and perceptions of faculty supervising final year undergraduates, specifically their perspectives of the supervision process, the different approaches taken to supervision and the challenges they face in supporting students through the dissertation journey.

While there are professional development opportunities for UG supervisors in Ireland and elsewhere, we have not encountered any that support the supervisor in designing, implementing and evaluating an OER artifact for their own supervisory practice. Claims for innovation are present in the literature in the form of collaborative and group-based supervision and there are instances of technology being used in UG supervision practice such as audio at the conclusion of supervisory meetings with recordings of students summarizing the discussion (Voelkel et al. 2018). However, supporting UG supervisors to consider their own supervisory style and the context of their practice before what for many for them is a new endeavor and places them outside their comfort zone in using a variety of multimedia tools is, we feel, a novel and engaging approach in this field.

Connecting with Open Educational Practices

An additional interest shared by the authors has been in emergent debates around open educational practices (OEPs) in higher education. We were keen to recognize these developments through the design of this module, and in the interests of developing and supporting supervisors of undergraduate research. Bates (2014) and Couros (2016) discuss the characteristics of the twenty-first-century educator in terms of openness and collaboration in practice, and creating, sharing and curating open educational resources (OERs) for teaching. In the context of this module, openness in teaching can be regarded as reflecting and discussing practice in the open, rather than in traditionally isolated or individual modes (Cronin and MacLaren 2018). In the creation and sharing of a multimedia resource, we encouraged participants to be open in terms of best practices with their peers and students. There was potential to share these resources more widely as OERs in their own right (Wiley 2015).

The Supervising Undergraduate Dissertations and Projects Module

The module was designed and validated in 2015, taking an expressly collaborative approach in its delivery and calling on practitioners to create resources that could support them in their work as supervisors. Table 9.1 shows the alignment of module learning outcomes, teaching and learning methods, and assessment strategy. Constructive alignment (Biggs 2003) was the theoretical underpinning of the outcomes-based module, with coherence between assessment, teaching and learning strategies and the intended learning outcomes. It was important that activities were designed that enabled participants to learn how to demonstrate achievement at the highest level described by the outcomes.

The module has been offered in the second semester of each academic year but one since 2015; 40 participants have now successfully completed it and ten participants have been students of our MA in Higher Education, working as faculty in TU Dublin or other HEIs across Ireland and in wide-ranging disciplinary contexts. These students had the option to take the module as an elective. The remainder have participated in the module on a stand-alone basis for continuing professional development, and have come from across the Colleges of TU Dublin:

- Arts and Tourism (13)
- Business (seven)
- Engineering and Built Environment (eight)
- Sciences and Health (two)

Table 9.1 Constructive alignment within the PD module

Learning outcomes	*Teaching and learning activities*	*Summative assessment*
Critically analyze what constitutes a productive undergraduate research learning environment. Explore conceptions of undergraduate research and supervisory practice, contextualized by critical engagement with salient and emergent issues in their own discipline.	Review of 'rethinking final year projects and dissertations: Creative honors and capstone projects' resource by Mick Healey—note thoughts on the introduction and select one of the case studies and discuss how it could be applied to own practice.	A reflective account, supported with reference to the literature, of the design and development of the resource, to include the context and underlying rationale as well as plans for implementation and consideration of the potential for wider application and dissemination.
Critically review the literature on the scholarship of undergraduate supervision pedagogy and of relevant policy issues in undergraduate research supervision. Evaluate their efficacy and competency in undergraduate research supervision.	Select one of the three provided journal articles and then summarize and critically analyze it in the online discussion board in the VLE.	
Discuss institutional requirements and procedures for undergraduate supervisors and research students, including ethics requirements.	Develop a question that could be sent in advance to the guest contributing in the final week. Participate in discussions with guest contributors.	Multimedia resource or resources (videos, screencasts with audio, word press site, infographic, etc.) that address two of the four supervision themes provided in relation to own context.
Evaluate and apply suitable undergraduate supervisory strategies and procedures for their own context.	As a learning set of four people, develop an overview of getting started/first steps in the undergraduate supervision process using a mind map in electronic format.	
Devise strategies for interactional and communication skills, e.g. negotiation, giving feedback, which is supportive and challenging.	Peer review session in final week—each participant speaks for 5 min about their resource and their peers and module leaders complete a short peer feedback form.	

The module was divided into five weekly workshops, focusing on specific themes: getting started with supervision; the identity and role of the UG supervisor; enhancing UG supervision practices to ensure impact (including the most effective ways of working and interacting with UG students in their dissertation); and disseminating good practice with input from other supervisors in the institution and beyond. As each cohort has a proportion of new supervisors, it was useful to explore self-perceptions of the novice supervisor's experience and attributes, issues affecting the novice supervisor's role, and supports and resources available for novice supervisors. One of the sessions took place in a computer room to allow participants to work on their OERs while tutors were present.

In terms of topics addressed in the module, a study in the Irish higher education context (Donnelly et al. 2013) revealed that supervisors identified the main student challenges in completing an undergraduate dissertation as pressure of work, managing time effectively and having the confidence needed for success. Deciding on a topic that was 'do-able' as well as knowing precisely what was expected at this level were also highlighted. The timing of the dissertation could also pose challenges, not least when undertaken by students with other modules in one semester. These areas were included as topics for exploration in the current curriculum for the PD module. Other themes explored are the culture of undergraduate research, supporting a program team approach to supervision, clarifying supervisor roles and student responsibilities, supporting the undergraduate dissertation process, exploring common issues in supervision, and assessing dissertations. When unpacking initial learning issues with the participants, topics that have emerged are GDPR impact on research (this is the EU General Data Protection Regulation, which was implemented in 2018 and is an important change in data privacy regulation), having a unified and agreed approach within their department or school to process (time management, research methods and clear procedures) and product of supervision (exploring opportunities for other alternate approaches to a dissertation/capstone project); how to manage new supervisors in the department; supervising across programs (what can be shared) and achieving consistency in feedback to students.

A key dimension of the module is the design and development of an authentic multimedia artifact by each participant to support their own supervision practice and framed around the themes explored in the module. To accompany this, participants are required to write a reflective and scholarly piece (Table 9.1). The multimedia resource can then be used as a resource by both students and academic faculty. Learners will have this additional support and guidance to help them as the resources deal with some of the common questions, concerns and practical issues that undergraduate students come across when completing their dissertation or final year project. The resource can also provide useful information for other faculty who are supervising undergraduate dissertations.

The aim is not to provide a set of definitive answers about supervising a dissertation or final-year project; instead participants will recognize that there are many ways in which the 'journey' through the supervision process can be completed. The resources can draw on a combination of the experiences of the dissertation supervisors on the module, academic research into faculty experiences of supervision, and examples of good practice.

Within the module workshops, input was invited from several guests who gave their perspectives on several approaches to undergraduate research implemented in our institution. The apprentice model was discussed as was a group research project. Also incorporated was an input on community-based research. This approach is applied in our university with the support of the Students Learning with Communities office. They have developed very clear guidance on what the roles and responsibilities of students and supervisors are in this context (Students Learning With Communities 2019).

Table 9.2 and Figs. 9.1, 9.2, 9.3, 9.4 and 9.5 give some examples of the range of multimedia artifacts in which participants chose to develop their most pressing supervision topics.

Methodology and Methods

We designed a short and focused evaluation of the module to address the following research question:

> *What is the perceived impact of sharing practices and creating multimedia artifacts in a professional development undergraduate supervision module in the context of the Irish higher education sector?*

All 40 graduates of the module were surveyed using an online questionnaire (Appendix A). The questions were developed through engagement with the literature on best practice in UG supervision as well as our

Table 9.2 Authentic multimedia artifacts on undergraduate supervision

	Type of multimedia resource	*Supervision topic*	*Supervision content*
2014–15 cohort	Video	Final-year group project support	Support for game development students during their final-year group project, addresses common problems that arise such as group conflict and the expectation that responsibility lies with students.
2015–16 cohort	Screencast	Checklist for submitting a group report (first years)	Assessment requirements for an enquiry-based group project report including Gantt charts.
2017–18 cohort	Infographic	Integrated learning portfolios	Structured guidance for students researching and compiling evidence of learning in a social care program
2018–19 cohort	Small-scale website	Academic writing and referencing	Tailored to the participant's discipline, a curated set of resources addressing writing and referencing from other well-regarded websites, with commentary.

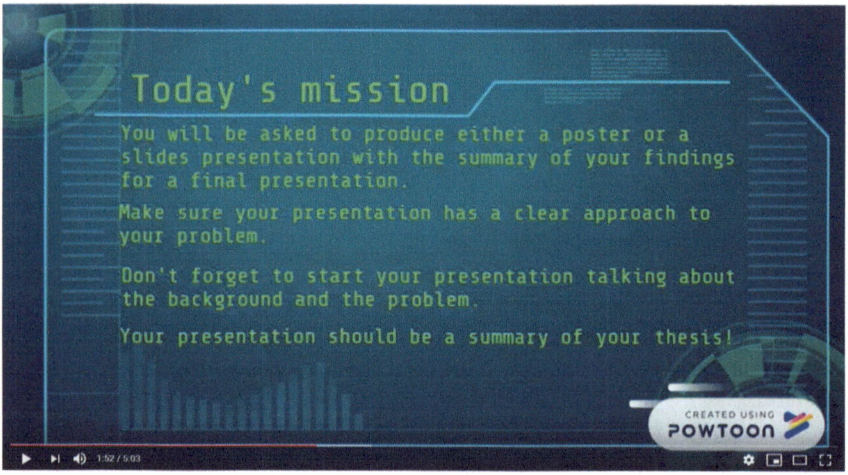

Fig. 9.1 Video resource for final year students by Camila D'Bastiani

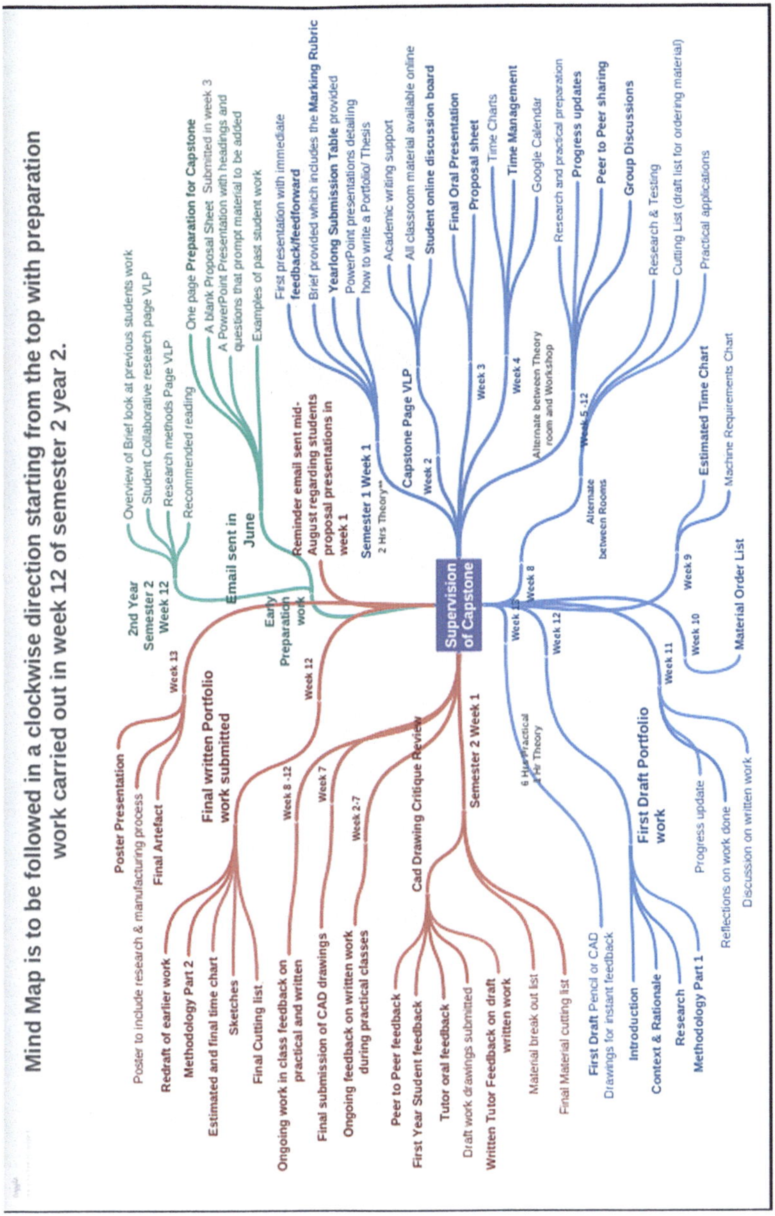

Fig. 9.2 Mindmap to support supervisors and students by Jennifer Byrne

Fig. 9.3 Infographic to support supervisors and students by Martina Ozonyia

Fig. 9.4 Video resource to support students by Michelle Bermingham

own expertise in delivering professional development in this area for a number of years.

There were three sections to the questionnaire, with the first asking closed questions to establish the profile of the participant—their discipline, current engagement with UG supervision, and the nature of the UG supervision taking place in their School/Institution. The second section focused on the UG Supervision module in relation to meeting the needs of UG supervisors in Ireland as a form of current professional development. It asked questions to establish participant motivations for undertaking the module as well as on the different areas of supervision that were considered important to each participant. Each was via provision of a 5-point Likert scale, followed by an open-ended question. The Likert scale was based on establishing importance in each instance it was used: the scale went from not important; somewhat; important; very; and not applicable.

Two open-ended questions were included to ascertain participant perceptions/understanding of what makes a good supervisor before they took the module, and after they had completed it. Further open questions were asked to build a picture of what makes an excellent undergraduate dissertation or research project, and what the supervisor can do to support production of that excellent work. A question on the impact of the

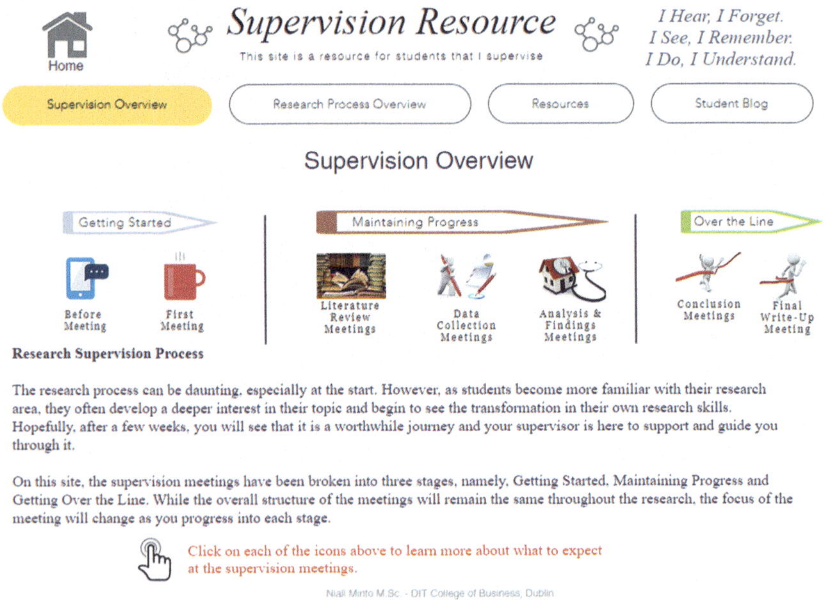

Fig. 9.5 Website to support students by Niall Minto

module was included on whether participants had continued researching resources or literature about undergraduate supervision. The third section of the online questionnaire asked open questions on participants' prior skillset with developing multimedia resources generally and in relation to supporting UG supervision, and a Likert scale was included to establish the extent of the impact that the multimedia artifact has had on students and colleagues in supporting the supervision process.

The Centre in which the module is offered has an existing approved protocol to address ethical issues in research relating to the evaluation of its programs, and we conducted the evaluation in line with this approved protocol. Seventeen people responded to the questionnaire; five from 2018 to 2019, seven from 2017 to 2018 and four from 2015 to 2016. All respondents were from TU Dublin—six respondents came from the College of Arts and Tourism, two from the College of Sciences and Health, five from the College of Business and three from the College of Engineering and the Built Environment. We present findings first in relation to participants' experiences of the module overall, and then in relation to the development of their multimedia resources.

Presentation and Discussion of Findings

Thirteen respondents were currently supervising UG dissertations, and two were not. From a logistical perspective, supervision of undergraduate research was undertaken as follows: 14 participants had supervision hours timetabled as part of their teaching schedule for meeting individual students for the duration of the supervision process; six participants had received submissions of draft work of the UG research project at different points in the semester in order to provide formative feedback to the students; three gave email advice and updates to their students; two provided online materials in the Webcourses/Brightspace VLE, and two used tutorials. The remainder was a mix of one participant who had hours included in their teaching timetable for group supervision, one having dedicated time in lectures or laboratory/practical classes, for example, for project management, and one using online submissions.

Rationale for Module Participation

Looking at the strongest reasons/motivations for doing the module: wanting to better support students in the supervision process was considered very important (12); closely followed by finding out about best practice in undergraduate research supervision (11); knowing how to deal with challenges in the supervision of undergraduate research (eight); meeting colleagues also engaged in undergraduate supervision (seven); and clarification of the supervisor's role in relation to research (six). Learning to develop a multimedia resource to support supervision was considered important/somewhat important by a total of 11 participants.

Participants were given the opportunity to share other reasons for undertaking the module and nine responded. There was a mix of wanting to learn from local practice, compare their own practices with colleagues, and explore the role of the supervisor in a supportive environment. Two were beginning to supervise dissertations, having had no previous experience in supervision, and three others wished to develop a consistent and fair supervision process, be trained how to supervise students properly and obtain clarification on standard practice and procedures around supervision. For one participant, this module formed part of the postgraduate qualification they were undertaking (MA in Higher Education).

Comparison of Attributes of a Good Supervisor Before and After the Module

It was interesting to note a set of attributes or characteristics of good practice in supervision that participants had before they undertook the module, and after it was completed. Figure 9.6 shows the combined similarities that appeared before and after the accredited professional development module, with advice based on analysis of the data shown in blue at the end of each section. These findings can be considered in the context to the work of Shanahan et al. (2015), who identified ten characteristics of effective undergraduate research mentorship. Seven of their characteristics are similar to the attributes below that emerged in this study. The remaining three do not feature in our findings and they relate to development of student mentoring skills, building a research community among students and supporting students in networking activities.

Perceptions of What Makes Excellent Undergraduate Research and the Role of the Supervisor to Support Excellent Work

Eight participants shared their perceptions of excellent undergraduate research, characterizing it in the following ways: 'engaged and interested' (Participant 5); 'literature review is linked to context and on to findings' (Participant 7); 'something unique and different' (Participant 10); 'succinct presentation of topic with a clear research question supported in the literature' (Participant 17); has a clear 'topic, methodology, theory and excellent writing' (Participant 11); 'that the student can demonstrate what they have learned from the project' (Participant 14); 'clear, concise work that answers the question' (Participant 16); 'good critical analysis of primary data linked with secondary data' (Participant 17), and 'a story from start to finish that adds to the research already there' (Participant 16).

The supervisor could contribute by: 'making sure the student was focused on key tasks' (Participant 7); supporting the student's decision-making on the research topic (Participant 9); 'guiding the student toward defining the aim/objective at the outset' (Participant 10); keeping in contact with the student and keeping the dissertation in line with the research questions (Participant 16); identifying the research question 'early on'

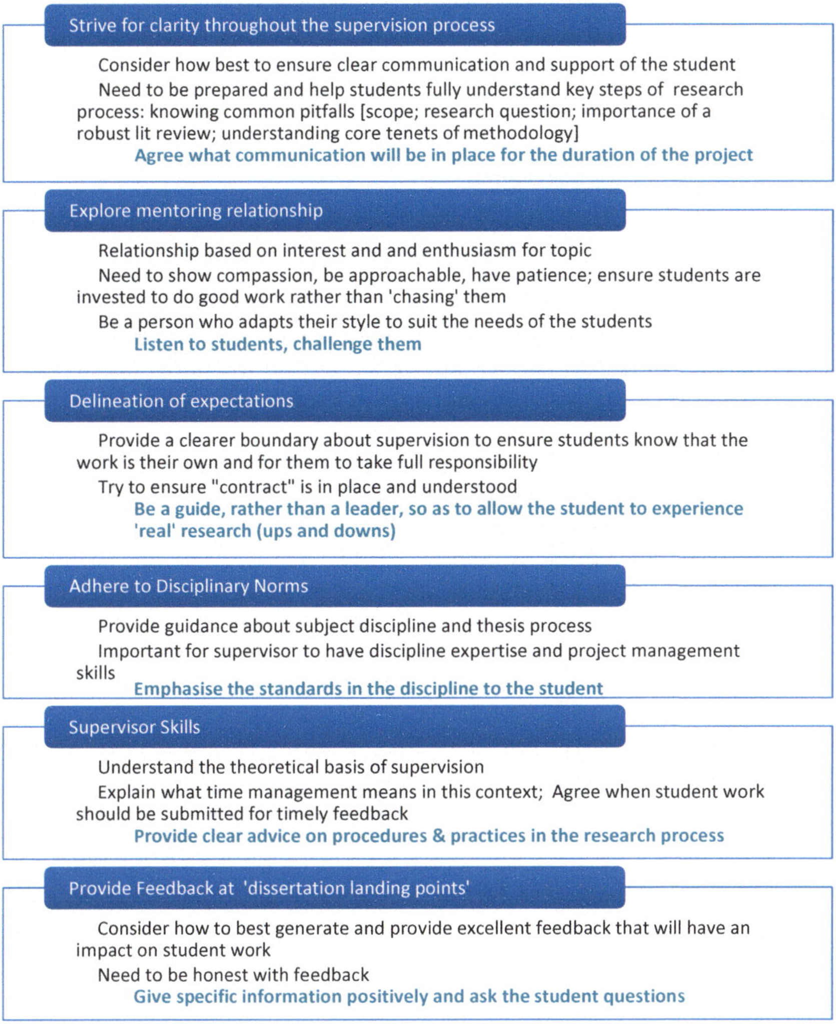

Fig. 9.6 Findings on good practice in undergraduate supervision

(Participant 17) so that 'an appropriate literature review would be undertaken, as many students spend most of their time on the literature review without having identified a good research question' (Participant 17). It can be useful to point supervisors to existing resources for supporting students to develop their research skills such as Willison and O'Regan's (2007) Research Skill Development Framework, which can be used to both chart and monitor students' research skill development. Conducting a research skills audit as part of the first supervisory meeting is an area that is discussed in the current module, and does not seem to form part of the existing practice of the participants.

Perceived Module Impact on Practice

Since completing the module, 11 participants have continued researching resources or literature about undergraduate supervision, and three indicated that they have not. Advice summarized from the participants' feedback on what they think should happen after the module included:

- Mentoring of Supervisors: 'Any new supervisors should shadow a qualified supervisor so they can learn from them' (Participant 7).
- Needs of New Supervisors: 'Staff should not undertake any supervision until this course is completed. [If] I was given the chance to design the course I would include more practical application for new supervisors e.g. how to complete a good literature review—understanding research methodologies and methods—common mistakes with quant/qual research' (Participant 7).
- Reassurance/Endurance: 'I found the module supported many of my current practices and that was reassuring. I also realized how I have to keep working at the process' (Participant 9).
- Network of Supervisors: 'Forming networks with other supervisors' (Participant 10); 'I have recommended it to many colleagues' (Participant 14).

Creation and Use of the Multimedia Resource

When asked if they had been using multimedia resources, technologies or apps to support supervision of undergraduate research before attending the module, 12 of the 17 respondents commented. Of these, five said that they had not been using such resources. One mentioned the resource created during the module. The remaining six were using a variety of resources: two had websites guiding students through supervision, one mentioned creation of YouTube clips for students, one used research papers, and one had a mix of material on the virtual learning environment including a separate Pinterest board for research in their module.

Before participating in the module, four people had created their own multimedia resources for students but ten others who responded had not. We asked participants in the research to comment on the creation and use of the multimedia resource by responding to a series of statements in a Likert-type question. We mixed positive and less positive statements about the process, to avoid leading the participants—14 people responded.

Most agreed or strongly agreed that it had been straightforward to create the multimedia resource, but two disagreed. Since the earlier iterations of the module, we have introduced a specific workshop to support the multimedia resource and this may have helped the more recent participants. Ten of the participants disagreed with the statement that it was difficult to think of a rationale for the resource, suggesting that most could think of a clear reason for developing it. Opinions were a little more mixed on the usefulness of the resource, nine agreeing or strongly agreeing that it had been useful and five disagreeing or not expressing a firm view on this. These findings reflect existing research on faculty use of open educational resources in Ireland (National Forum 2015), which show a similar pattern of somewhat uneven use of online resources, and caution amongst faculty around creating, using and reusing resources.

Most of the participants said they would be happy to share the resource with colleagues in their department or School, with just one out of 14 indicating they were not sure about this question. However, responses to sharing beyond the institution were more mixed: 11 people said they would agree they would be happy to do this, three were less certain. Again, this may reflect a more general wariness around sharing educational resources in the Irish higher education sector (National Forum 2015). This is something we would like to address further through the module since the sharing of practice has been of central importance to participants, and they have indicated the value of this. This in turn would support greater openness in practice and pedagogy with the creation and sharing of resources (Cronin and MacLaren 2018). The early iterations of the module emphasized copyright issues and also used the feedback process to comment on specific issues for each resource, something we have continued to do in the most recent cohorts.

We presented a statement suggesting that the multimedia resource had benefited students undertaking undergraduate research. It could be anticipated that this might be difficult to answer unless participants had evaluated the use of the resource by their students. The responses appeared to indicate this with eight people agreeing or strongly agreeing, but one choosing 'not applicable' and five neither agreeing nor disagreeing with the statement. It was perhaps also difficult for them to comment on the extent to which the resource benefited their colleagues supervising undergraduate students: five agreed or agreed strongly with this, five neither agreed nor disagreed; with two saying it was not applicable to them and two disagreeing. Within the module, we have not given space to discussion of evaluation of the resource and therefore it is likely that participants are not seeking feedback from their students about the value of using the resources. When asked to respond to a statement about gathering formal feedback about the resource, just three people said that they had done this. This is an area that we could address much further in future, and it may also be appropriate to incorporate students' contributions to further resources in keeping with a student-as-partner approach to curriculum (Healey et al. 2014).

Finally, participants were asked to respond to a statement about whether they would like to create further resources for their students. Seven people said that they would, but six did not commit to a view on this and one person disagreed with the statement. This is perhaps a little disappointing since the module offered scope to open up this possibility and perhaps support the creation of suites of resources within the different disciplines. With an opportunity to explore this further, we will ask participants whether they used their multimedia skills elsewhere on completion of the module in other professional projects. This could be an aspect of practice that we seek to develop more fully in future iterations. Later, when asked if they had created further resources, five people indicated that they had. These included ongoing development of a website, creation of a separate set of resources and guides, finding existing third-party materials and sharing them online, and formalizing processes through creation of forms to support supervision. Two participants commented that they did not have time to create further resources.

It appears that a follow-up workshop for all participants in the module next year would be warranted, to facilitate participants in making changes before they might choose to release their resources publicly. They could also be given advice as to how to evaluate the resources being used, pitch new OERs and potentially collaborate with each other in producing these.

Macro Level Issues

From the findings, it is interesting to consider the bigger picture issues discussed in this book. As we are based in a newly established Technological University in Ireland, the first of its kind nationally, the scope of undergraduate inquiry and research needs to be more visible and supported. There are currently some inter-institutional undergraduate research celebration days (http://sure-network.ie/about/; https://www.ul.ie/ctl/students/all-ireland-conference-undergraduate-research-aicur), and this can be built upon as can students-as-partners in research (faculty and students coauthoring papers and copresenting at regional and national conferences). Development of enquiry and research skills earlier within the curriculum is also an important consideration (Healey et al. 2013).

Conclusions and Recommendations

In this chapter, we have presented the arguments for greater professional development opportunities for faculty supervising undergraduate research, contextualizing this within the Irish higher education sector. We described and explained our professional development module, which is available to faculty undertaking supervision of undergraduate research at a new Technological University, TU Dublin as well as faculty based in other Irish HEIs. Our evaluation of this module showed that participants had explored and articulated the characteristics of good, even excellent, undergraduate research. They perceived a positive impact on their practice from having had the opportunity to do this through the module by talking with their peers and sharing practice. They identified a range of pathways toward successful completion of the research dissertation or project, and this has been reflected in the wide-ranging multimedia resources developed over the past four years. While at this stage it might be somewhat premature to talk of the module overtly in terms of OEPs, and the resources as OERs, this valuable perspective offers us several directions in which to develop the work in future. We suggest that there needs to be growing recognition of the importance of undergraduate research, and that it should be celebrated in Ireland more widely in the ways we have seen happening internationally. This would in turn raise the standing of good undergraduate supervision, and recognize the efforts and supports discussed by colleagues in this short module. We will continue to develop and evaluate this module over the longer term, potentially through the development of mentoring and networks of support for new supervisors within the disciplines. We would also like to encourage greater sharing of the multimedia resources given the participants' time and effort invested in producing these. Our findings throughout this research have repeatedly shown the value and importance of collegial discussion in building confidence and resilience amongst faculty meeting the needs of larger and ever more diverse groups of students (Higher Education Authority 2018). We conclude by encouraging colleagues nationally and internationally to address support for supervisors through appropriate PD, and particularly through allowing critical conversations amongst colleagues to take place in supportive spaces.

Appendix A: Evaluation Questionnaire

Section A: The Basics

In which year did you complete the Supervising Undergraduate Dissertations and Projects module? [Select from list of 2019, 2018, 2016, 2015].

In which College of the City Campus are you based? [Select from Sciences and Health, Business, Engineering and Built Environment, Arts and Tourism].

Are you currently engaged in supervising undergraduate research? [Select from Yes/No].

How is supervision of undergraduate research currently undertaken in your School? (Please tick any applicable) [List included below]

- Supervision hours timetabled for individual students
- Supervision hours timetabled for group supervision
- Dedicated time in lectures or practicals, e.g. for project management, work-in-progress, updates
- Email advice and updates
- Provision of online materials in Webcourses/Brightspace
- Submissions of draft work at different points in the semester
- Tutorials
- Using an online discussion board or forum

Section B: The Module

Were any of the following important in your decision to take the Supervising Undergraduate Dissertations and Projects module? [Likert Scale question using scale: not important; somewhat important; important; very important; not applicable]

- Clarification of the supervisor's role in relation to research
- Finding out about best practice in undergraduate research supervision
- Learning to develop a multimedia resource to support supervision
- Meeting colleagues also engaged in undergraduate supervision
- Knowing how to deal with challenges in the supervision of undergraduate research
- Wanting to better support your students in the supervision process

Were there any other reasons to take the module? [Open text response].

What did you think were the attributes of a good supervisor *before you took the module*? [Open text response].

What do you think are the attributes of a good supervisor *having completed the module*? [Open text response].

Can you comment on what makes an excellent undergraduate dissertation or research project, and what the Supervisor can do to support production of that excellent work? [Open text response].

Since completing the Supervising Undergraduate Dissertations and Projects module, have you continued researching resources or literature about undergraduate supervision? [Select Yes/No].

Section C: The Multimedia Resource

Can you give any examples of how you were using multimedia resources, technologies, or apps to support supervision of undergraduate research before you attended the Supervising Undergraduate Dissertations and Projects module? [Open text response].

Had you ever created your own multimedia resource for your students before participating in the Supervising Undergraduate Dissertations and Projects module? [Select Yes/No].

Please indicate your responses to the following statements: [Likert Scale question using scale disagree strongly; disagree; neither agree nor disagree; agree; agree strongly; not applicable]

- I found it fairly straightforward to create the multimedia resource
- It was difficult to think of a rationale for the multimedia resource
- The multimedia resource has been useful in my supervision of undergraduate research
- I had a clear idea of a typical student in mind when I designed the multimedia resource
- I am happy to share the multimedia resource with colleagues in my department/School
- I would be happy to share my multimedia resource with any colleague internally or externally to the institution
- The multimedia resource has benefited the students undertaking undergraduate research
- The multimedia resource has benefited my colleagues supervising undergraduate students
- I have gathered formal feedback about the multimedia resource from my students and/or colleagues
- I would like to create more multimedia resources for my students from now on

Since completing the Supervising Undergraduate Dissertations and Projects module, have you produced other resources for your colleagues or students? [Select Yes/No].

Can you give any further details in relation to the previous question?

Section D: Conclusion

Do you have any further comments in relation to the module or the questions raised by this questionnaire?

References

Baker, M., Cluett, E., Ireland, L., Reading, S., & Rourke, S. (2014). Supervising undergraduate research: A collective approach utilising group-work and peer support. *Nurse Education Today, 34*(4), 637–642.

Bates, S. (2014). *Anatomy of 21st century educators.* Retrieved from: https://www.slideshare.net/EdPER_talks/the-anatomy-of-the-21st-century-educator

Biggs, J. (2003). *Teaching for quality learning at university – What the student does* (2nd ed.). Buckingham: SRHE/Open University Press.

Bologna Working Group. (2005). *A framework for qualifications of the European Higher Education Area*. Bologna Working Group Report on Qualifications Frameworks, Danish Ministry of Science, Technology and Innovation.

Boud, D. (1999). Situating academic development in professional work: Using peer learning. *The International Journal for Academic Development, 4*(1), 3–10.

Couros, G. (2016). *10 essential characteristics of a 21st century educator. Connected principals sharing. Learning. Leading*. Retrieved from: http://connectedprincipals.com/archives/13542

Cronin, C., & MacLaren, I. (2018). Conceptualising OEP: A review of theoretical and empirical literature in open educational practices. *Open Praxis, 10*(2), 127–143. Retrieved from https://files.eric.ed.gov/fulltext/EJ1177676.pdf.

Donnelly, R., Dallat, J., & Fitzmaurice, M. (Eds.). (2013). *Supervising and writing a good undergraduate dissertation*. Oak Park: Bentham Science Publishers.

Grant, J., Schofield, M. J., & Crawford, S. (2012). Managing difficulties in supervision: Supervisors' perspectives. *Journal of Counseling Psychology, 59*(4), 528–541.

Hammick, M., & Acker, S. (1998). Undergraduate research supervision: A gender analysis. *Studies in Higher Education, 23*(3), 335–347.

Hanratty, O., Higgs, B., & Tan, E. (2011). Irish perspectives on undergraduate research. *Council on Undergraduate Research Quarterly, 31*(4), 33–42.

Healey, M., & Jenkins, A. (2018). The role of academic developers in embedding high-impact undergraduate research and inquiry in mainstream higher education: Twenty years' reflection. *International Journal for Academic Development, 23*(1), 52–64.

Healey, M., Lannin, L., Stibble, A., & Derounin, J. (2013). *Developing and enhancing undergraduate final-year projects and dissertations*. Retrieved from https://www.heacademy.ac.uk/knowledge-hub/developing-and-enhancing-undergraduate-final-year-projects-and-dissertations

Healey, M., Flint, A., & Harrington, K. (2014). *Engagement through partnership: Students as partners in learning and teaching in higher education*. York: Higher Education Academy.

Higher Education Authority. (2017). *National framework for doctoral education*. Dublin: HEA.

Higher Education Authority. (2018). *Key facts and figures: Higher education 2017/8*. Dublin: HEA. Retrieved from https://hea.ie/assets/uploads/2019/01/Higher-Education-Authority-Key-Facts-Figures-2017-18.pdf

Larson, S., Partridge, L., Walkington, H., Wuetherick, B., & Moore, J. L. (2018). An international conversation about mentored undergraduate research and inquiry and academic development. *International Journal for Academic Development, 23*(1), 6–14.

Lee, A. (2012). *Successful research supervision: Advising students doing research.* London: Routledge.

Moore, J. L., & Felten, P. (2018). Academic development in support of mentored undergraduate research and inquiry. *International Journal for Academic Development, 23*(1), 1–5.

NAIRTL. (2011). *NAIRTL grants initiative: Evaluation of impact.* Cork: NAIRTL.

NAIRTL. (2012). *Developing an institutional framework for supporting supervisors of research students: A practical guide.* Cork: NAIRTL.

National Forum for the Enhancement of Teaching and Learning in Higher Education. (2015). *Learning resources and open access in higher education institutions in Ireland.* Retrieved from https://www.teachingandlearning.ie/publication/learning-resources-and-open-access-in-higher-education-institutions-in-ireland/

National Forum for the Enhancement of Teaching and Learning in Higher Education. (2016). *The national professional development framework.* Retrieved from https://www.teachingandlearning.ie/publication/national-professional-development-framework-for-all-staff-who-teach-in-higher-education/

National Qualifications Authority of Ireland (NQAI). (2003). *National framework of qualifications.* Dublin: NQAI.

Quality and Qualifications Ireland. (2006). *Verification of compatibility of Irish national framework of qualifications with the framework for qualifications of the European higher education area.* Retrieved from https://www.qqi.ie/Publications/Publications/Verification%20of%20Compatibility%20of%20NFQ%20with%20QF%20EHEA%20220609.pdf

Roberts, L. D., & Seaman, K. (2017). Good undergraduate dissertation supervision: Perspectives of supervisors and dissertation coordinators. *IJAD, 23*(1), 28–40.

Roberts, L. D., & Seaman, K. (2018). Students' experiences of undergraduate dissertation supervision. *Frontiers in Education, 3*, 109.

Rowley, J. (2000). Thirteen tips for successful supervision of undergraduate dissertations. *Educational Developments, 1*(1), 14–15.

Rowley, J., & Slack, F. (2004). What is the future for undergraduate dissertations? *Education and Training, 46*(4), 176–181.

Shanahan, J. O., Ackley-Holbrook, E., Hall, E., Stewart, K., & Walkington, H. (2015). Ten salient practices of undergraduate research mentors: A review of the literature. *Mentoring & Tutoring: Partnership in Learning, 23*(5), 359–376.

Students Learning With Communities. (2019). *Guidelines on roles for students community partners and lecturers.* Retrieved from http://www.dit.ie/media/ace/slwc/worddocuments/TU%20Dublin%20guidelines%20on%20roles%20for%20students,%20supervisors%20and%20community%20partners.pdf

The Guardian, Academics Anonymous. (2017, July 28). Not all PhD supervisors are natural mentors – Some need training. *The Guardian*. Retrieved from: https://www.theguardian.com/higher-education-network/2017/jul/28/not-all-phd-supervisors-are-natural-mentors-some-need-training?CMP=new_1194

Todd, M. J., Smith, K., & Bannister, P. (2006). Supervising a social science undergraduate dissertation: Staff experiences and perceptions. *Teaching in Higher Education, 11*(2), 161–173.

UK Council for Graduate Education. (2019). *The good supervisory practice framework*. Retrieved from: https://supervision.ukcge.ac.uk/good-supervisory-practice-framework/

Vereijken, M. W. (2017). Novice supervisors' practices and dilemmatic space in supervision of student research projects. *Teaching in Higher Education, 23*(4).

Voelkel, S., Mello, L.V., & Varga-Atkin, T. (2018). Supporting students during their undergraduate research projects using audio recordings. *Innovations in Education and Teaching International, 55*(4), 433–440.

Warhurst, R. P. (2006). "We Really Felt Part of Something": Participatory learning among peers within a university teaching-development community of practice. *International Journal for Academic Development, 11*(2), 111–122.

Wiley, D. (2015) 'Defining the "Open" in Open Content and Open Educational Resources' [Online]. Available at http://opencontent.org/definition/ (Accessed 4 March 2019).

Willison, J., & O'Regan, K. (2007). Commonly known, commonly not known, totally unknown: A framework for students becoming researchers. *Higher Education Research and Development, 26*(4), 393–409.

Wisker, G. (2012). *The Good Supervisor: Supervising Postgraduate and Undergraduate Research for Doctoral Theses and Dissertations*. London: Palgrave Macmillan.

Wynaden, D., Wichmann, H., & Murray, S. (2013). A synopsis of the mental health concerns of university students: Results of a text-based online survey from one Australian university. *Higher Education Research & Development, 32*(5), 846–860.

Zydney, A.L., Bennett, J.S., Shahid, A., & Bauer, K. (2002). Faculty perspectives regarding the undergraduate research experience in science and engineering. *Journal of Engineering Education, 91*(3), 291–297.

CHAPTER 10

Research in Transforming Contexts: Ensuring Relevance and Impact

Mandla S. Makhanya

Introduction and Background

Traditionally, research universities are dedicated generators of new knowledge and PhDs, deemed to be vital for the social and economic development of any country. However, the ongoing influence and impact of technology as a driver of the knowledge economy (and the 4th Industrial Revolution) and the increasing and cumulative sophistication of its various applications across an array of fields has contributed to a growing link between, and focus on, STEM research in particular.

This has given rise to a relatively small and select, but highly influential group of universities whose funding streams are diverse, whose focus and reach are global and whose aim is to produce innovative, cutting-edge research. Mohrman et al. (2007:1) named these universities *Emerging Global Model* (EGM) research institutions, which are characterized by an intensity of cutting-edge research, as well as worldwide competition for students, faculty, staff, and funding. EGM institutions are at the vanguard

M. S. Makhanya (✉)
University of South Africa, Pretoria, South Africa
e-mail: makhams@unisa.ac.za

N. H. Hensel, P. Blessinger (eds.), *International Perspectives on Undergraduate Research*,
https://doi.org/10.1007/978-3-030-53559-9_10

of research that is driving managed socioeconomic development and progress at a global scale. Furthermore, by aligning the role of the research university with the neoliberal view of education as a marketable commodity, EGM Universities have created a powerful dynamic around the role and influence of research as a directed, saleable commodity. Although relatively few, these are the institutions that head virtually every list of leading universities worldwide and include universities such as Harvard, Cambridge, Stanford, California-Berkley, MIT, Caltech, Columbia, Princeton, Chicago, Oxford, Yale and Cornell.

Table 10.1 reflects the top 20 universities globally, as derived from a study by Times Higher Education and QS Top Universities (2019), and clearly demonstrates the preponderance of these influential universities in the northern hemisphere. This trend is supported and to a large extent echoed by the other ranking instruments such as the QS World University rankings, which similarly compares and evaluates universities based on six quantitative criteria, with the main emphasis being on research (Table 10.2).

The rankings indicate overwhelmingly that most "top" universities are in North America with the remainder in the United Kingdom and Zurich, Switzerland. While universities in China have made strong showings in recent years and now have their own ranking systems (QS 2019) significant disparities are evidenced in ranking criteria and performance on these, placing them on the periphery of the most influential global universities. Africa and South Africa do not currently have their own ranking instruments either, but feature in both the QS and the THE education rankings (Table 10.3).

The QS 2019 ranked the top universities in Africa as follows:

1. University of Cape Town
2. The American University in Cairo
3. University of the Witwatersrand
4. Stellenbosch University
5. University of Johannesburg
6. University of Pretoria
7. Ain Shams University
8. Assuit University

The ranking of higher education institutions remains highly contentious although their ongoing popularity and use in branding and

Table 10.1 Times Higher Education (THE) World University Rankings 2019—Top 20

Rank	*Name*	*Overall*	*Teaching*	*Research*	*Citations*	*Industry income*	*International outlook*
1	University of Oxford United Kingdom	96.0	91.8	99.5	99.1	67.0	96.3
2	University of Cambridge United Kingdom	94.8	92.1	98.8	97.1	52.9	94.3
3	Stanford University United States	94.7	93.6	96.8	99.9	64.6	79.3
4	Massachusetts Institute of Technology United States	94.2	91.9	92.7	99.9	87.6	89.0
5	California Institute of Technology United States	94.1	94.5	97.2	99.2	88.2	62.3
6	Harvard University United States	93.6	90.1	98.4	99.6	48.7	79.7
7	Princeton University United States	92.3	89.9	93.6	99.4	57.3	80.1
8	Yale University United States	91.3	91.6	93.5	97.8	51.5	68.3
9	Imperial College London United Kingdom	90.3	85.8	87.7	97.8	67.3	97.1
10	University of Chicago United States	90.2	90.2	90.1	99.0	41.4	70.9
11	ETH Zurich Switzerland	89.3	83.3	91.4	93.8	56.1	98.2
=12	Johns Hopkins University United States	89.0	81.9	90.5	98.5	95.5	71.9

(*continued*)

Table 10.1 (continued)

Rank	*Name*	*Overall*	*Teaching*	*Research*	*Citations*	*Industry income*	*International outlook*
=12	University of Pennsylvania United States	89.0	87.4	89.2	98.4	70.3	63.6
14	UCL United Kingdom	87.8	79.1	90.1	95.9	42.4	95.8
15	University of California, Berkeley United States	87.7	78.7	92.3	99.7	49.3	69.8
16	Columbia University United States	87.2	85.4	83.1	98.8	44.8	79.0
17	University of California, Los Angeles United States	86.4	82.6	87.9	97.8	49.4	62.1
18	Duke University United States	85.4	84.1	78.8	98.2	100.0	61.0
19	Cornell University United States	85.1	79.7	85.4	97.4	36.9	71.8
20	Arbor United	84.1	80.0	85.9	96.0	45.9	58.0

marketing speaks to the generally perceived and accepted status of the rankings as a measure of excellence and performance amongst higher education institutions. Perhaps more crucially though, the rankings serve to highlight the significant divide between the so-called North and the South, as well as the difficulty faced by the vast majority of other universities, in competing in any truly meaningful way, against these top performers. This view becomes increasingly relevant where developmental needs, available resources, and complex socioeconomic and political dynamics circumscribe the scope and content of research (if at all) at universities in developing nations. This view is supported by Badat (2010:1) who questions the notion of university rankings and their relevance for the Global South and alludes to their inherent hegemonic influences and impacts.

Table 10.2 QS World University Rankings 2020 and 2019—Top 10

2020 rank	*2019 rank*	*University*	*Location*
1	1	Massachusetts Institute of Technology (MIT)	United States
2	2	Stanford University	United States
3	3	Harvard University	United States
4	5	University of Oxford	United Kingdom
5	4	California Institute of Technology (Caltech)	United States
6	7	ETH Zurich (Swiss Federal Institute of Technology)	Switzerland
7	6	University of Cambridge	United Kingdom
8	10	UCL (University College London)	United Kingdom
9	8	Imperial College London	United Kingdom
10	9	University of Chicago	United States

Table 10.3 Times Higher Education (THE) World University Rankings 2018 edition ranked top South African universities

SA rank	*World rank*	*University*
1	171[4]	University of Cape Town
2	251–300[5]	University of the Witwatersrand
3	351–400[6]	Stellenbosch University
4	401–500[7]	University of KwaZulu-Natal
5	601–800[8]	University of Pretoria
6	601–800[9]	University of Johannesburg
7	601–800[10]	University of the Western Cape
8	801–1000[11]	University of South Africa

In addition, virtually all the research conducted by these institutions tends to employ scientific methods of enquiry, even in disciplines outside of the sciences. Geiger cited in Mohrman et al. (2007:146) asserts that at the heart of the EGM is an expansion of the older functions of teaching, research, and service into an organization that can best be described as a "knowledge conglomerate." Increasingly this view is coming into contestation with discernable shifts to Multi-, Inter-, and Transdisciplinarity (MIT) research that encourages the harnessing of a variety of methodologies across a number of disciplines, using a multiplicity of skills and training, in pursuit of more broadly relevant and applicable research.

Pau (2003:151) explains that much of the funding for this type of research comes from the private sector and a large number of the research projects are in fact done in partnership with business, to the extent that many EGM institutions, through their partnerships and funding, conduct specific types of research for their funders alone. This suggests that in this model, research has moved beyond the goal of knowledge acquisition in the purer academic sense, to the commissioned acquisition of specified kinds of knowledge for the purpose of exploiting their global applicability and marketability. It can therefore be argued that in such cases, the end goal of the research would appear to be primarily, the generation of income.

Interestingly, this profit-generation dynamic is increasingly evident in non-EGM universities across the globe as the need for third-stream income from innovation, patents and business generation takes hold in the context of funding constraints and a general global economic downturn, which is manifesting amongst others, in decreased income from state funding and other traditional income streams for higher education institutions.

Altbach (2006:151) summarizes these competing typologies and tensions, asserting the prestige and dominance of "scientific" research over non-"scientific" fields of enquiry; the advantage of English speakers over non-English speakers when it comes to accessing the most prestigious research publications; the inferior status of teaching in relation to research in the institutional hierarchy; the scorning of more esoteric disciplines and research initiatives in favor of the more practical, fund-generating initiatives that find favor with research partners and government; the inability of nations or institutions with limited financial resources to compete in the very expensive research "game"; the stagnation or decline in currency of higher education institutions or campuses that are not currently research-intensive; the preservation of language and culture not being seen as competitive with those who are discovering new knowledge; the necessity of accepting the methods, norms, and values of the universities in Western Europe and North America that currently dominate the system in order to join the international marketplace of ideas, especially in science; and the adherence to established research paradigms, irrespective of whether or not the themes and subject areas of interest to leading scientists are relevant to universities at the periphery.

These observations have become increasingly apposite, especially on the African Continent and South Africa in particular (and in other developing nations), as the striving for the decolonization of education and research is gaining ground in most universities.

There is evidently a widening gulf, based on historical advantage, wealth and privilege, between the top research institutions and the rest, which is not only broadening the research divide between North and South, but also impeding development and the achievement of the SDGs in historically disadvantaged nations. This is substantiated by the fact that numerous studies place research outputs from (for example) Africa, at less than 2% of global research outputs, with the major portion of that research emanating from South African Universities. It is also asserted that a significant portion of the research is in collaboration with scholars in the North, so it is currently quite difficult to determine the precise percentage of research done by individual African researchers (Tijjsen and Winnick 2018; Le Roux 2015). At the same time, the continued dominance of STEM research is contributing to the declining status of other fields such as the humanities and social sciences, underscoring the assertions of knowledge hegemony advanced by Ball (2006:15) and Badat (1999:17).

The foregoing discussion demonstrates quite clearly that research in developing nations and in Africa and South Africa in particular, requires urgent and innovative interventions and strategizing if it is to grow and develop, ensure progress on the SDGs and national development, and assert its status on the global stage. Furthermore, the narrow focus on STEM research needs to be addressed in the developing context, given that present and future socioeconomic and political challenges require research that is focused on humanities and social sciences and not only on STEM research. As Nowotny et al. (2001) assert: "There is an urgent need for a new mandate for science if it is to deal more effectively with complex societal challenges."

It is precisely this current global state of flux and fundamental reorganization, and the incongruent and disintegrated sources of knowledge and information across all spheres of human endeavor, which call for a more thoughtful and nuanced contemplation of the future of research in higher education.

The Transforming Higher Education Context

The current transformation in higher education is so profound that it is calling into question the very notion of the university and compelling us to adjust in fundamental ways to ensure our survival and our sustainability. Key drivers of this transformation include:

- Rapid technological innovation and advancement undergirded by expanding digitization and (inter)connectivity, the current phase of which is dubbed the 4th Industrial Revolution (4IR)
- Global concerns around the sustainability of the planet and its people and linked to the achievement of the SDGs and notions of global citizenship, social justice, inclusiveness and fairness
- An increasingly complex, interdependent, borderless, and deeply unequal world
- A growing array of role players/stakeholders/participants in the global higher education arena who are leveraging technological and digital affordances to promote their agendas and who are producing new research and knowledge outside of the bounds of the university

Research finds itself at the nexus of these drivers, immersed in an ever-widening, compounded cycle of innovation and response. The limits of the innovation are unknown and the speed of its uptake and implementation are unprecedented and unparalleled, begging the question: What key factors should we be considering in our design of models and strategies to ensure our relevance and impact in transforming contexts? Three key factors will be considered below.

Niche Research Focus Areas/Themes Aligned to National Socioeconomic and Developmental Priorities

Given the limited financial resources available to most universities nowadays (including research-intensive universities), as well as institutional pressures on university leadership to ensure a material Return On Investment (ROI), especially where public funding is involved, the relative luxury of "research for research" sake has largely disappeared (Tarran 2010).

In South Africa, Africa and many other developmental societies, the combined realities of a demand for ROI and the need to ensure socioeconomic development and relevance has resulted in a clear alignment of institutional research strategies with national and continental policy and developmental needs. These are supported by the establishment of various national, regional, and continental research associations and bodies whose aim is to integrate or align education policy, and promote quality, relevant scholarship, including research that has at its heart, social justice and development. It will therefore come as no surprise to note that research focus areas or themes in many developing nations are closely aligned to their national developmental and social justice needs, even as they aim to be globally relevant.

In South Africa, such legislation and policy in support of national development include:

- The Higher Education Amendment Act, No. 9 of 2016
- White Paper for Post-School Education and Training: Building an Expanded, Effective and Integrated Post-School System (2013)
- Education White Paper: A Programme for Higher Education Transformation (1997)
- National Skills Development Strategy III, DHET (2011)
- National Skills Development Plan (2018)
- Human Resource Development Strategy of South Africa (2010–2030) (2010)
- National Development Plan: Vision for 2030 (2013)
- The Sustainable Development Goals (2016)
- The UN Agenda 2063: The Africa We Want (2015)

The University of South Africa's Research & Innovation Strategic Plan (2019–2020) echoes those of many African universities in terms of its focus and intention and as such, Unisa has been used as an example of current research sentiments and trends in South Africa and on the Continent. Unisa is a dedicated Open Distance and eLearning (ODeL) university, (the largest on the Continent) and while it is not a research-intensive University, it nevertheless has a very dynamic Research &

Innovation portfolio, whose ambitious strategic intent articulates the belief that its own success in research, innovation, and postgraduate studies has significant potential to influence the future of research in South Africa and in Africa as a whole. In this belief, the portfolio also invests in, and supports, postgraduate studies, research, innovation, and commercialization projects and partnerships across the institution, by fostering collaborations within and across Unisa colleges, and with research councils, other universities, private sector entities, and government departments.

Echoing this ethos, the following research themes/focus areas linked to national development needs may be discerned in some of the foremost (South) African universities, the majority of which appear in the global rankings mentioned earlier. Clearly, universities in developmental contexts have realized the imperative to be both relevant and competitive in their own right and it is exciting to note a number of cutting-edge projects that have been receiving global and continental recognition and acclaim for their excellence, relevance, applicability, and contribution to the continental body of knowledge.

The Joint Science Academies' statement: *Science and technology for African development* asserts:

> African countries must be able to develop, adapt and exploit scientific and technological solutions appropriate to their specific needs, otherwise they risk becoming ever more dependent on advice and assistance from the developed world……Without embedding science, technology and innovation in development we fear that ambitions for Africa will fail.

It would, however, be unfortunate to adopt a view of research that focuses narrowly on STEM and marginalizes Arts, Human and Social Sciences. It is becoming increasingly evident that the fundamental social reorganization that is taking place as technology advances and transforms our world will necessitate a vast amount of so-called soft discipline research to anticipate and propose innovative and pragmatic means of dealing with the social disruption and dynamics that have already begun occurring. It is in this context that the need for multi-, inter-, and transciplinary research comes to the fore.

Targeted Development of Research, Especially Multi-, Inter-, and Transciplinary Research (MIT) Toward More Integrated and Relevant Research Outcomes and Application

Interdisciplinary Research

The National Science Foundation in the USA defines Interdisciplinary Research as follows: "Interdisciplinary research is a mode of research by teams or individuals that integrates information, data, techniques, tools, perspectives, concepts, and/or theories from two or more disciplines or bodies of specialized knowledge to advance fundamental understanding or to solve problems whose solutions are beyond the scope of a single discipline or area of research practice." In some instances, interdisciplinary research leads to knowledge being integrated in such a way that a field beyond the original disciplines is created, often referred to as transdisciplinary research. It is evident though that there are broad conceptualizations, definitions, and interpretations of both interdisciplinary and transdisciplinary research and that these frequently overlap. The nature of the research conducted will inevitably offer the clearest evidence of whether the research is interdisciplinary, transdisciplinary, or both.

Multi-, Inter-, and Transdisciplinarity (MIT)

Multi-, Inter-, and Transdisciplinary (MIT) research is perhaps less known but concerns that which is at once between the disciplines, across the different disciplines, and beyond all disciplines. Its goal is the understanding of the present world, of which one of the imperatives is the unity of knowledge. The other imperative is the generation of knowledge that has transformative heuristics. The openness inherent in transdisciplinarity involves an acceptance of the unknown, the unexpected, and the unforeseeable; while the tolerance inbuilt within it implies acknowledging the right to ideas and truths opposed to our own. MIT thus forms a departure from the more traditional understandings and practices of research to pave the way for new knowledges, understanding, and practices that have particular relevance and application for the societies and countries in which they are found and from which they emanate.

MIT can be implemented to great effect as has been demonstrated at the University of South Africa, which is a key proponent of this methodology. Considering the projects mentioned subsequently, it is evident that most, if not all universities in South Africa and many in Africa, have embraced conceptualizations of MIT in research practice.

MIT is fundamental to Unisa's research strategy and practice. This ethos is evident in many developmental contexts and is supported in most cases by policy. The university uses several goals that are shared with other universities:

1. Collaborations with African universities and other centers of knowledge and promoting interactions with intellectuals and thinkers on the continent and Diaspora with a view to addressing critical issues relating to Africa's situation in the contemporary global context and commit to realigning education and its relevance to society in Africa
2. Promoting the recognition of indigenous knowledge throughout the continent through conscientious and context-sensitive research and sharing of insights emanating from this knowledge system
3. Restructuring relations with community holders of knowledge as fellow experts in the generation of knowledge. Research and development strategies must integrate ethical considerations as well as issues of the protection of Intellectual Property Rights, economic benefit sharing, poverty alleviation, and employment creation. Key tenets and aspects of IKS must be systematized and integrated into the curriculum
4. Proactively promoting and supporting MIT as a distinct and strategic institutional approach to addressing Africa's developmental imperatives and working toward developing and implementing support and incentivization strategies for transdisciplinary initiatives, including flagship initiatives that infuse MIT methodologies and approaches to research and curriculum development and practice

To this end, Unisa has a number of Chairs and is involved in a variety of projects, often in collaboration with other institutions and researchers across disciplines and at other universities. Unisa has already completed, or is engaged in, a number of initiatives that are not only exemplars of research excellence, but also deeply transformative in line with the promotion of indigenous knowledge, MIT, and the SDGs. These include:

- *MIT postgraduate studies including at Masters and PhD* levels across most disciplines. The defining characteristic of these qualifications is that the student will be required to demonstrate high-level research capability and make a significant and original academic contribution at the frontiers of the discipline or field. This work must be of a quality to satisfy peer review and merit publication. There is an array of niche areas across all the faculties that are contextually relevant, and which are calculated to dovetail with one another for research purposes.
- *Peace and Human Development Chair: Cultural Resources for Peace Building*

  The Chair takes up the research subject of peace and human development in Africa as a means of introducing critical perspectives on democracy, values, jurisprudence, human rights and human wrongs and the place of responsibility of different cultures, including peace building from an African perspective. The issue of peace, conflict resolution, peacebuilding, and recently, restorative justice is an area-cluster that can consolidate transdisciplinarity as an approach to discourse, practice, and thought.
- *Science, Culture, and Society*: *Science, Plurality, and Other Ways of Seeing*

  This research area takes the pronouncements contained in the UNESCO Declaration on Science for the Twenty-First Century, which states that all cultures can contribute scientific knowledge of universal value, and that there is, therefore, a need for a vigorous, informed, and democratic debate on the production and use of scientific knowledge. In order to help find ways of better linking modern science to the broader heritage of humankind, the Chair undertakes deep analyses of the linkages between science in relation to cosmology, constitution, citizenship, community, and syllabi, thus making propositions for curriculum reform and transformation.
- *Indigenous Knowledge Systems and Innovations: The Conditions for their Integration.* In the context of this Chair, Indigenous Knowledge is seen as part of the subaltern and heterogeneous forms of knowledge that had no place in the fields of knowledge that grew in compact with colonialism and science. Theoretically, Indigenous Knowledge Systems makes it possible to explore mean-

ings and theories of death, obsolescence, resilience, survival, globalization, freedom, and healing. It enables us to revisit concepts like "property" and the "commons" as well as the systems that govern these concepts. By taking on IKS, the Chair contemplates questions such as: What are the possibilities for alternative globalizations, alternative regimes of intellectual property, and of alternative times?

- *Universities and Society: Rethinking Community Engagement.* The Chair engages in the articulation of issues lying at the interface between university and society in Africa; and thus invest in cultivating a theory of praxis through linkages with innovative nonformal centers/indigenous communities in Africa and internationally with the aim of generating new insights and building discourse coalitions on the transformation of universities within South Africa, Africa, and beyond.

A number of community-focused initiatives prove the value and impact of MIT in tackling social challenges. These are directly linked to the SDGs and demonstrate the power of alternative approaches to sharing and leveraging knowledge and capacities.

- *Fog Harvesting Project*

A climatologist at Unisa's School of Agriculture and Environmental Sciences helped develop a system to harvest moisture from abundant mountain fog in a water-scarce region of the Eastern Cape. Funding came from the Water Research Commission, and the fog water system was designed in collaboration with colleagues from Pretoria University. Unisa is involved in ongoing research into water harvesting from fog, especially for isolated rural communities. The project has also been rolled out in other dry areas of South Africa. Villagers' lives have changed with the installation of the water-harvesting system and its inexhaustible supply. No electricity is needed to power the scheme, which makes it eco-friendly and low-cost, and suitable for areas with no power infrastructure.

- *Biogas Project*

Unisa's Exxaro Chair in Business and Climate Change (Exxaro Chair), the South African National Energy Development Institute (SANEDI) and the University of Fort Hare (UFH), are rolling out biogas as a renewable and sustainable source of energy. Researchers drawn from three colleges at Unisa (College of Economic and Management Sciences, College of Science, Engineering and Technology, and the College of Agriculture and Environmental Sciences) are involved in a transdisciplinary, interdisciplinary, and multidisciplinary research programs focusing on bio-derived fuels (BDF) and solar technology transition under climate change and the green economy. Biogas is one of the project streams in the research program. The project witnessed the installation of 13 household biogas digesters that will generate cooking gas. Waste from cattle, goats, and pigs is being used as feedstock. The by-product (digestate slurry) is a very good source of fertilizer (rich in nitrates and phosphates), which can be used for growing vegetables, fruits, and flowers.

- *The Institute for Dispute Resolution (IDRA)* is located within the College of Law.

However, its research agenda is not confined to the legal discourse. IDRA is a multi/inter/transdisciplinary institution (MIT) that transcends the borders of demarcated academic disciplines and building bridges for the gaps and dissonance formed/forming by sociopolitical conditions of our continent's past century by means of community-engaged research. IDRA is also in conversation with those continents to which we are closely related, be it politically, economically, or spiritually, and with whom we can share knowledge systems of dispute processing and dispute resolution. Researchers come from: the legal profession; the legal academy; English literature studies; social work; political sciences; and anthropology, and they share language, ideas, theory, experience, and spirit to create a team of researchers who reflect the diversity of the research institute. IDRA currently has postdoctoral fellows, masters, and doctorate students from countries such as the Congo, Ethiopia, Cameroon, and South Africa. The postdoctoral fellows are engaged in field and desktop research to build and evolve the literature in conflict, peace, and alternate dispute resolution. The research agenda is focused on developing a body of knowledge based on humanistic values, such as the values of *Ubuntu*, which promotes a harmony model of dispute resolution. The research agenda fosters sensitivity to the cultural context of African community spaces, be they local or diasporic. The research agenda strives to serve humanity. A number of Unisa's Colleges also include IKS in the recurriculation of their courses.

- *Transdisciplinary African Psychologies programmes (TAP)*

TAP understands African Psychologies as covering all areas that mainstream Western Psychology covers, but from an African-situated decolonizing approach, as well as areas not investigated, or neglected by Western Psychology. As a transdisciplinary decolonizing program, TAP is interested in identifying points of convergence between disciplines of psychology and those that study Africa and Africans. TAP is the formalization of a space to bring about a dedicated and specific Africa-centered study of the intersectionalities related to power, race, identity gender, violence, community, the collective psyche, and much more. It builds on the traditions of criticality, compassion, and centeredness that the Institute has come to be known for over the past three decades.

These are some of the exciting projects in MIT, not only at Unisa but evidently at a number of South African and African universities. Examples of exciting medical breakthroughs in Africa centered medical challenges, using collaboration, MIT, harnessing cutting-edge and innovation technology and techniques have been noted at most of the research-intensive universities identified in the THE and QS rankings. Some of these include: *Wits scientists closer to slowing the progressions of Alzheimers*; *SA doctors cure deafness using 3D printing tech* (University of Pretoria medical team); and *UCT team in immune system breakthrough*. All these research projects point to the growing ubiquity of MIT as a key foundation stone of teaching and learning, research, and community engagement and as the key driver of relevant knowledge production. Shared ownership of knowledge and knowledge generation, strategies, efforts, and acknowledgment are sometimes a difficult barrier to overcome but in Africa this is made a lot easier by the spirit of Ubuntu, collectivism, and community that characterizes most of African society.

Given the unique challenges presented in developmental contexts, especially in regard to a chronic lack of resources and capacities, it should come as no surprise that MIT research is delivering such exciting results in South Africa and in many other developing nations.

Developing New Cohorts of Researchers

One of the key challenges for research in developing contexts is the limited number of researchers being produced. This is exacerbated by the flight of many qualified researchers to the developed nations, for reasons that range from political whimsy, a lack of resources, nonsupportive work environments, and in many instances, antipathy or a lack of support for women in research. If research is to be taken seriously in developmental contexts, then all of these must be addressed.

Understanding the need and urgency to address these inhibiting factors, a growing number of governments and universities, especially in South Africa, are putting in place dedicated strategies, and making resources available to incentivize and grow new, young researchers who will remain in their institutions or on the continent and contribute to national and continental growth and development. Most, if not all these emerging researchers will have had first-hand experience and understanding of their geo- social, economic, and political challenges, possibly lending much needed nuanced appreciation and depth to the contextualization of their research and the ranking and focus of their research endeavors.

Underpinning all of these strategies remain the globally sanctioned and practiced commitments to integrity, quality, rigor, and ethics in research; the intention to ensure high-quality researchers and research capacities; the cultivation and promotion of institutional ethos, intellectual cultures, and research experiences that are conducive to critical discourse, intellectual curiosity, tolerance, and a diversity of views; contributing to society by producing Master's and Doctoral graduates of sound character, versatile ability, and knowledge; and meeting the research needs of the globally competitive society by nurturing collaborative relationships with its stakeholders and other partners.

Thus, many universities now have strategies and plans in place to grow researchers, increase research outputs, employ more effective and relevant research methodologies, and practices, devise focus areas or themes that harness their "natural" strengths, including those linked to their locations and the biological and agricultural resources (See Table 10.4 above) and most importantly, a dedicated program to support and incentivize the development of academics as researchers. Some of these initiatives include:

Table 10.4 Research focus/themes in (South) African universities, indicating geo-environmental and indigenous foci and developmental priorities

Research focus/themes in (South) African universities, indicating geo-environmental and indigenous foci and developmental priorities						
University of Pretoria (University of Pretoria. https://www.up.ac.za/en/research-innovation/article/257889/research-focus-areas)	*University of Cape Town (University of Cape Town. https://www.uct.ac.za/sites/default/files/image_tool/images/328/research/publications/reports/UCT%20Research%20Strategy%2015%20November%202014.pdf)*	*University of the Witwatersrand (Wits University. http://www.wits.ac.za/research/research-a%2D%2D-z/)*	*Stellenbosch University (Stellenbosch University. http://www.sun.ac.za/english/research-innovation)*	*University of Johannesburg (University of Johannesburg. https://www.uj.ac.za/research/)*	*American University of Cairo (American University of Cairo. https://www.aucegypt.edu/research/research-centers)*	*University of South Africa. (An ODeL University) (University of South Africa. Research & Innovation Strategic Plan. 2019–2020. University of South Africa. Not for public consumption)*
Translational medicine (human and animal disease)	Minerals to Metals	Health	Social Justice and development	African Art	Yousef Jameel Science and Technology Research Center	Knowledge generation and human capital development in response to the needs of South Africa and the African continent
African Wildlife Health and Management	Drug Development	Food	Health and human security	Artificial intelligence and machine learning	Prince Alwaleed Bin Talal Alsaud Centre for American Studies and Research	The promotion of democracy, human rights and responsible citizenship

Pathobiology of Disease	Brain-Behaviour Initiative (BBI)	Sustainable water management: agriculture, effluent management, filtration, and water in society	Systems and technologies for the future	Astronomy and Cosmology	Cynthia Nelson Institute for Gender and Women's Studies	Innovation and capacity building in science and technology
Sustainable livelihoods and Well-being	Marine Research Institute (MA-RE),	Engineering	Human creativity and social innovation	Cities	Access to Knowledge for Development Centre (A2K4D)	Economic and environmental sustainability
	Cities in Africa	Microbiology	The natural environment	Communicable diseases	The Centre for Translation Studies (CTS)	ODeL
		Polymer science and agriculture		Data Science	Middle East Studies Centre (MESC)	

(*continued*)

Table 10.4 (continued)

Research focus/themes in (South) African universities, indicating geo-environmental and indigenous foci and developmental priorities

University of Pretoria (University of Pretoria. https://www.up.ac.za/en/research-innovation/article/257889/research-focus-areas)	*University of Cape Town (University of Cape Town. https://www.uct.ac.za/sites/default/files/image_tool/images/328/research/publications/reports/UCT%20Research%20Strategy%2015%20November%202014.pdf\)*	*University of the Witwatersrand (Wits University. http://www.wits.ac.za/research/research-a%2D%2D-z/)*	*Stellenbosch University (Stellenbosch University. http://www.sun.ac.za/english/research-innovation)*	*University of Johannesburg (University of Johannesburg. https://www.uj.ac.za/research/)*	*American University of Cairo (American University of Cairo. https://www.aucegypt.edu/research/research-centers)*	*University of South Africa. (An ODeL University) (University of South Africa. Research & Innovation Strategic Plan. 2019–2020. University of South Africa. Not for public consumption)*
		Conservation ecology		Drug delivery	Kamal Adham Centre for Television and Digital Journalism	
		Private law		Equality studies in the Global South setting	Centre for Migration and Refugee Studies (CMRS)	
		Health sciences		Global change and sustainability	John D. Gerhart Centre for Philanthropy and Civic Engagement and Responsible Business	

Philosophy	Mathematics and mathematical educated	El Khazendar Business Research and Case Centre
	Geoscience and mining engineering	Social Research Centre
	Noncommunicable diseases	AUC Forum
	Paleoanthropology, Paleontology and Palynology	Centre for Applied Research on the Environment and Sustainability
	Refugee and migrant studies	
	Material science and engineering	

- Research opportunities extended beyond the academy to include professional and academic staff. This is particularly relevant in ODeL where the pedagogical enterprise is totally contingent on the transactional and operational environments.
- Aggressive development of PhD students and (rated) researchers through writing workshops, mentoring, exposure to conferences and the delivery of conference papers, support in applications for rankings and aligned to that, individual research plans to ensure progression in publication and rankings.
- Acknowledgments and incentivization for excellence in research through research awards dinners, financial awards, generous research grants and awards, and ad hominem promotions for research productivity and progress.
- Programs to increase the number of doctoral staff, for example, through fully paid "sabbaticals" with job retention, conditional on the successful completion of the PhD in the allotted time.
- Regular National Research Youth Conferences; institutional research weeks aimed at showcasing and sharing with peers, research progress, developments, achievements and innovations; research symposia and workshops in collaboration with other institutions and government, all of which are aimed at growing and retaining the cohort of researchers. For example, The South African Tertiary Network recently partnered with the Department of Higher Education and Training to launch a Staff PhD Capacity Enhancement Programme designed to raise the number and quality of PhDs coming out of universities of technology and previously disadvantaged universities in South Africa. The program aims to give 50 aspiring PhD students across 11 South African Universities of Technology and previously disadvantaged universities the opportunity to complete their studies with the help of top lecturers and professors in the country and abroad. The program aims to reduce the completion times and dropout rates of PhD candidates in South Africa, which is currently an estimated 60%.
- A commitment to advancing women in research, and especially black women, through targeted support and incentivization including in regard to career progression and placement.

Fundamental to this approach is a commitment to excellence and support for early career researchers, emerging research leaders, and researchers from previously disadvantaged designated groups. Given the urgent need for the development of researchers if we are to meet and overcome the challenges posed by developmental states and the SDGs and the imperative for genuine and lasting transformation, this approach to research development is, quite simply, the most pragmatic and ethical thing to do.

There is already evidence that this approach is bearing fruit. In a recent keynote address to the Unisa research community, Dr David Green (Unisa 2018), Digital Resources Director of the Taylor & Francis Group, stated the following:

> Based on total Web of Science papers, there were four times as many articles by African authors in 2016 compared to 2000, and a 280% increase in the number of South African authors in this timeframe...[Green said] the continent's share of global research had increased from 2.31% in 2010 to 3.24% in 2016...."South Africa is now on 0.99%," [adding that] after Egypt, the country was the biggest contributor in Africa, followed by Algeria, Nigeria and Kenya....

However, where South Africa is coming into its own is in the number of citations its researchers receive—citations being "a rough proxy" for quality content. In 2016, the global average for citations per article was just over 1.1. South Africa is a fraction away from this, and ahead of China, Japan, Brazil, India, and Russia. "South Africa is very close to the world average and you will soon overtake that, I am sure. However black female researchers remain underrepresented at only 14% of the current research cohort."

Referring to the impact of technology on research publications, Green referred to the growing phenomenon of online, open access formats stating [that] "In the United Kingdom, 37% of research is now open access and China is up to 40%; around 25% of South African research is now published open access—the global average." (2018).

Conclusion

It remains somehow ironic that research, as a transformative force for socioeconomic development and the good of ordinary citizens, continues to be underappreciated and underresourced in many developing contexts. It has been demonstrated though, that by focusing on the three key strategic areas for research development in developmental contexts, outlines above, demonstrable progress can be made in ensuring its relevance and impact and in lending impetus to genuine transformation for good.

References

Altbach, P. G. (2006). Globalization and the university: Realities in an unequal world. In Forest, J.J.F. and Altbach, P.G., Reisberg, L.E. 2009. *Trends in global higher education: tracking an academic revolution.* Executive Summary. A Report Prepared for the UNESCO 2009.

Badat, S. (1999). *Black student politics, higher education and apartheid: From SASO to SANSCO, 1968–1990.* Pretoria: Human Sciences Research Council.

Badat, S. (2010, April 14). *Have global university rankings any value for the global South ? Universities in crisis.* Blog of the International Sociological Association. Available at: http://www.isa-sociology.org/universities-in-crisis/?p=421

Ball, S. J. (2006). *Education policy and social class: The selected works of Stephen J. Ball (World Library of Education Series).* London: Routledge.

Department of Higher Education and Training (DHET). *Human resource development strategy for South Africa. A nation at work for a better life for all.* Available at http://www.dhet.gov.za/Reports%20Doc%20Library/Human%20Resource%20Development%20Strategy%20for%20South%20Africa.pdf.

Green, D. (2018). Keynote Address to Unisa Community in 2018. Dr David Green, Digital Resources Director, Taylor & Francis Group.

Le Roux, E. (2015). Publishing South African scholarship in the global academic community. *Notes and Records of the Royal Society of London, 69*(3), 301–320. https://doi.org/10.1098/rsnr.2015.0033.

Mohrman, K., Wanhua, M., & Baker, D. (2007). The emerging global model of the research university in higher education in the new century. In P. G. Althbach & P. M. Peterson (Eds.), *Global challenges and innovative ideas.* Rotterdam: Sense Publishers.

Nowotny, H., Scott, P., & Gibbons, M. (2001). Re-thinking science: Knowledge and the public in an age of uncertainty. *Contemporary Sociology, 32*. https://doi.org/10.2307/3089636.

Pau, B. (2003). Cited in Mohrman K, Wanhua M & Baker, D. 2007. The emerging global model of the research university in higher education in the new century. In P. G. Altbach & P. M. Peterson (Eds.), *Global challenges and innovative ideas* (p. 2007). Rotterdam: Sense Publishers.

QS Top universities in Africa. Available at: https://www.topuniversities.com/university-rankings-articles/world-university-rankings/top-universities-africa

QS World University Rankings. *Top universities in China in 2019*. Available at: https://www.topuniversities.com/university-rankings-articles/mainland-china/top-universities-china-2019

QS World University Rankings. *2020 and 2019 – Top 10*. Available at: https://www.topuniversities.com/university-rankings/world-university-rankings/2019

Republic of South Africa (RSA). (2011). *National skills development strategy III*. Department of Higher Education and Training. Available at: https://www.gov.za/sites/default/files/gcis_document/201409/national-skills-development-strategyiiia.pdf

Republic of South Africa (RSA). (2012). *National development plan 2030*. Available at: https://www.gov.za/issues/national-development-plan-2030

Republic of South Africa (RSA). (2013). *Education white paper 3. A programme for the transformation of higher education*. Pretoria: Department of Education.

Republic of South Africa (RSA). (2016, January 17). *The Higher Education Amendment Act, No 9 of 2016*. Government Gazette. Available at: https://www.gov.za/sites/default/files/gcis_document/201701/40548act9of2016higheducationamendmentact2016_0.pdf

Republic of South Africa (RSA). (2018). *National skills development plan*. Available at: http://www.dhet.gov.za/Gazette/12%20January%202018%20THE%20NATIONAL%20SKILLS%20DEVELOPMENT%20PLAN%20(NSDP).pdf

Republic of South Africa. (RSA). (2013). *White paper for post-school education and training: Building an expanded, effective and integrated post-school system*. Department of Higher Education and Training. Republic of South Africa. http://www.dhet.gov.za/SiteAssets/Latest%20News/White%20paper%20for%20post-school%20education%20and%20training.pdf

Rhodes University News. (2013). *Researchers unite for SA policy push*. Available at: https://www.ru.ac.za/perspective/latestnews/2013archive/researchersunite-forsapolicypush.html

Tarran, B. (2010). 2010 predictions – Research for research's sake is over, says Forrester. *Research Live*. Available at: https://www.research-live.com/article/news/2010-predictions%2D%2Dresearch-for-researchs-sake-is-over-says-foresters/id/4001786

The Joint Science Academies'. *Science and technology for African development*. Available at: http://www.interacademies.net/File.aspx?id=6431 (Undated).

The Sustainable Development Goals. (2016). Available at: https://www.undp.org/content/undp/en/home/sustainable-development-goals.html

Tijjsen, R., & Winnick, J. (2018). *What does Africa contribute to global research?* Available at: http://www0.sun.ac.za/scistip/news/what-does-africa-contribute/

Times Higher Education (THE) World University Rankings 2019: Top 20. Times Higher Education World University Rankings (2019). Accessed at: https://www.timeshighereducation.com/world-university-rankings/2019/subject-

Times Higher Education World University Rankings. (2018). Accessed at: https://www.timeshighereducation.com/world-university-rankings/2019/subject-

Times Higher Education World University Rankings. *Top universities in South Africa*. Available at: https://www.timeshighereducation.com/student/where-to-study/study-in-south-africa/ (Undated).

United Nations (UN) Agenda 2063. *The Africa We Want*. (2015). Available at: https://www.un.org/en/africa/osaa/pdf/au/agenda2063.pdf

University of Cape Town. (2018). *UCT team in immune system breakthrough*. https://www.news.uct.ac.za/article/-2018-12-21-uct-team-in-immune-system-breakthrough

University of Pretoria (UP). (2019). *Academic pioneers world's first middle ear transplant using 3d printed bones*. https://www.up.ac.za/news/post_2750323-up-academic-pioneers-worlds-first-middle-ear-transplant-using-3d-printed-bones

University of South Africa. (2013). *Profile of the DST/NRF South African Research Chair in Development Education at the University of South Africa*. Available at: https://www.unisa.ac.za/static/corporate_web/Content/Research%20&%20innovation/Documents/SARChI_Chair_Development_hoppers.pdf

University of South Africa. (2018). *How Africa is growing its share of global research*. Available at: https://www.unisa.ac.za/sites/corporate/default/Research-&-Innovation/News-&-events/Articles/Copy-of-How-Africa-is-growing-its-share-of-global-research

University of South Africa. (UNISA). Research & Innovation Strategic Plan. 2019–2020. University of South Africa. Not for public consumption.

Wits University. (2018). *Wits scientists closer to slowing the progressions of Alzheimers*. Wits University. https://www.wits.ac.za/news/latest-news/research-news/2018/2018-09/wits-scientis-closer-to-slowing-down-progression-of-alzheimers.html

CHAPTER 11

Undergraduate Research in France

Isabelle Mirbel and Margarida Romero

France, a National Ambition for an Excellence-Oriented Research

France has been historically recognized among key players in Higher Education. Nevertheless, the last three decades has challenged the Higher Education context not only by a growing internationalization of the programs (Bedenlier et al. 2018), but also by the rise of Higher Education in emerging countries (Yang 2018) and distance learning initiatives worldwide (Garrett 2019). Among the objectives of the French Ministry of Higher Education, Research, and Innovation (*Ministère de l'Enseignement supérieur, de la Recherche et de l'Innovation, MESRI*), research is an important focus. In the French government, Frédérique Vidal, Minister of

I. Mirbel (✉)
Laboratoire d'Informatique, Signaux et Systèmes de Sophia-Antipolis (I3S) – UMR7271 – UNS CNRS, Université Côte d'Azur, Nice, France
e-mail: Isabelle.Mirbel@univ-cotedazur.fr

M. Romero
Laboratoire d'Innovation Numerique pour l'Education, Université Côte d'Azur, Nice, France
e-mail: margarida.romero@fse.ulaval.ca

N. H. Hensel, P. Blessinger (eds.), *International Perspectives on Undergraduate Research*,
https://doi.org/10.1007/978-3-030-53559-9_11

Higher Education, Research, and Innovation, aims to "bring together research, innovation and economic clusters" through "an ambitious national roadmap and a strong involvement in the Horizon 2020 European program" (Open Access Government 2019). The MESRI created the Initiative of Excellence (IDEX programs) aiming to create intensive research universities in France. Traditionally, France has had a centralized decision-making process in Higher Education. However, a decentralization process started in 2007 with the adoption of the Liberties and Responsibilities of Universities law, referred to as the LRU (after the acronym of the official name *Loi relative aux libertés et Responsabilités des Universités*), gave universities more autonomy in the management of their resources but also regarding their research strategies. After making the LRU, the IDEX programs have developed further the decentralization and internationalization of French universities by funding specific IDEX programs of selected universities. IDEX programs have played an important role in creating new opportunities for undergraduate research, which we will describe later in this chapter.

A Research System Combining National Research Centers and University Research Units

France is oriented toward building a strong institutional system concerning research. It includes a diversity of research institutions counting national research centers, research labs within the universities' structures, and joint research units (UMR) governed by the largest public research organization CNRS (*Centre National de la Recherche Scientifique*) and universities. Most of the researchers in France are affiliated to universities while carrying out a teaching and research career. In the following section, we describe undergraduate research in national institutes before focusing on the different undergraduate research initiatives at the French university system.

Undergraduate Research in National Institutes

CNRS is the most renowned and important research center at the national level; researchers are organized into disciplinary research units. In addition to CNRS, there are other national institutes with specific expertise, such as the expertise in computer sciences (INRIA) or agriculture (INRA). A total of 36 national institutes have a specific expertise in different fields, and specific ways to interact within their field (MESRI 2018). Within these various structures, undergraduate research has not been organized nationally but can be observed through specific initiatives allowing undergraduates to be engaged in some research activities. For instance, the Center for Molecular Biophysics (CBM), a research unit of the French National Center for Scientific Research (CNRS), hosts undergraduate trainees in biochemistry, biology, physics, chemistry, and computer sciences in collaboration with the University of Orléans. The initiatives to support undergraduate research are not coordinated at the national level. Therefore, the state of undergraduate research significantly differs among the different national institutes of research. Nevertheless, its existence highly depends on the individual initiatives of the research institutes and researchers working within these structures.

Undergraduate Research in the French Universities

Undergraduate research is defined as the inquiry or investigation conducted by an undergraduate student that makes an original intellectual or creative contribution to the discipline (Altman et al. 2019). In France, the MESRI does not provide the official definition of the undergraduate research and the use of the term is rare. Based on research within search engines and the French scientific database, CAIRN, there is only one initiative that has been explicitly classified as "undergraduate research". The undergraduate research initiative was developed between France and Kosovo in the field of Mathematics and the study was conducted in French-Canadian universities in Québec (Chouinard et al. 2015). The "France-Kosovo Undergraduate Research School of Mathematics" (*Collège franco-kosovar de formation à la recherche en mathématiques*) brought together "the most gifted students in mathematics from the University of Pristina and young French mathematicians for nine intensive lectures devoted to modern fields of mathematics" in 2017 (Milyon Lab 2017).

The lack of an official definition reflects that the term undergraduate research is not being used in France at the current moment. Traditionally, research is not considered as the main objective of undergraduate curricula. An introduction to research for specific disciplines in France has often been considered in the form of a course at the master level, but not among the main objectives of undergraduate programs. Undergraduate studies in France are not commonly associated with the acquisition of research competencies. Undergraduate students are expected to develop a scientific-based knowledge and competencies in their graduate courses, mostly taught by faculty with a research load in addition to their teaching responsibilities. In some cases, the research labs propose undergraduate fellowships for students to take part in various research projects. The tradition of research assistants is well established, for example, in archeology and other social sciences, but less common in other areas such as learning sciences. Finally, as to our knowledge, "community-based research" and citizen science are still not officially structured at the undergraduate level in France although they are becoming increasingly popular internationally, especially in the field of environmental studies.

Undergraduate programs are not always about research units for developing joint activities. Partnerships between undergraduate studies and research labs are not common in France except in the case of the international research labs that already have a tradition to host undergraduate research fellowships. The "apprentice model" for undergraduate research in France does not have a clear status for students aiming to secure an apprenticeship in a research lab. However, a new form of "student contract" has been recently developed with the aim of giving a possibility to students to engage in diverse projects, including research.

The French government leads the initiatives to encourage research and innovation through the MESRI. From one side, there is a growing number of initiatives to encourage the improvement of Higher Education quality in terms of learning strategies, especially related to competency-based education and the development of new programs («Nouveaux cursus à l'université» within the *Programme d'investissements d'avenir*, PIA). Other initiatives aim to develop entrepreneurship in Higher Education through the development of undergraduate and graduate startups (e.g. Pépite program). The development of undergraduate research is not included in these different programs but has been considered by some universities supported by the IDEX program, such as University of Côte d'Azur or University of Paris Descartes. Within the Université Côte d'Azur, the Invent@UCA program encourages interdisciplinary projects using a design thinking approach. The Invent@UCA initiative also helped undergraduate and graduate students to develop their entrepreneurship skills (Cassone et al. 2019). Most of the existing initiatives in undergraduate research have been developed under the umbrella of the IDEX program. Within a quality and excellence orientation in Higher Education,

MESRI proposed new funding programs to support the reform of Higher Education: Initiatives of Excellence (IDEX) programs. The purpose of these programs is to create world-class multidisciplinary higher education and research institutions in France. Although most universities do not offer undergraduate research in all disciplinary fields, there are a growing number of initiatives within several universities. Within the IDEX undergraduate initiatives supporting research, we will make a distinction between (a) university-level initiatives aiming to support undergraduate research across the various disciplines and for all the students at the university, (b) program-based initiatives, and (c) course-based initiatives.

University-Level Initiatives for Supporting Undergraduate Research

In this section, we introduce four programs developed at the university-level, which are made available for students enrolled in the different undergraduate programs at UPEC, Paris-Descartes, Paris 8, Lille University, and Bordeaux University. Among the IDEX-funded universities, Université Paris-Est Créteil Val de Marne (UPEC) has developed initiatives to encourage undergraduate and graduate students to develop their research competencies. In the context of the Council of Studies and University Life (*Conseil des Études et de la Vie étudiante*, CEVU), the UPEC has created the Scientific council for the introduction of research in the undergraduate and masters' levels (*Conseil Scientifique sur l'initiation et la formation à la recherche en licence et en master*). The scientific council has been created to encourage research training as part of the undergraduate and masters' programs (Regnaut 2012). To our knowledge, no other French universities have created official steps to develop undergraduate research. Beyond the UPEC initiative, there is not an organization or government agency at MESRI that supports undergraduate research. In most cases, the objective of this kind of training is to introduce undergraduates to research and potentially develop their interest in scientific careers by preparing them for masters and doctorate programs.

The second initiative of undergraduate research at the university level is developed by Paris-Descartes University. Paris-Descartes has developed a call for projects (2019) for supporting faculty lead initiatives in undergraduate research. Université Paris-Descartes "wishes to strengthen the link between training and research and allow students to learn about the practice of research in the second or third year of their undergraduate course. This initiation will take the form of internships in laboratories and/or participation in congresses, colloquia or seminars, and will be supplemented by explicit elements of methodology (e.g. writing of research reports in English, ethics in research, introduction to scientific controversies ...)". The call for projects is oriented toward the faculty community, which is invited to develop a coherent proposal within the context of their faculty.

A third university-level initiative is developed at Paris 8 University. The distance program *Think, Wonder, Problematize, and Evaluate* ("Penser, s'Etonner, Problématiser et Evaluer [PEPE])" aims at introducing research to undergraduate students from different fields and helping them to develop critical thinking competencies. The courses are personalized according to the interests of undergraduate students and allow them to develop their research competencies. The curriculum of the PEPE program is presented in Fig. 11.1.

Through the evaluation of PEPE program Meunier and Zibetti (2019) observed that the undergraduate students find this training as necessary preparation for masters' studies. The study of Meunier and Zibetti showed the students who attended the PEPE program developed competencies for reflective and critical analysis of documents on chosen research topics.

Lille University developed the fourth initiative of undergraduate research. This northern France university created an elective course aiming to develop an introduction to research. The course could be validated by doing an internship in a research lab of approximately 10 days. The introduction to the research course is available for undergraduate students

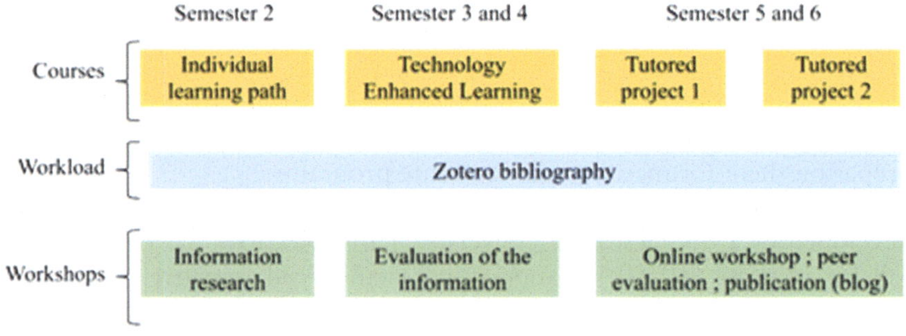

Fig. 11.1 Organization of the PEPE program

of all the disciplines available at Lille University: Sciences and Technology, Economic Sciences and Management and Social and Human Sciences.

The efforts to develop research competencies of undergraduate students were incorporated at the Bordeaux University when they engaged in the pedagogical transformation of the university. One of the objectives of the university is to integrate research training of students within the existing courses and pedagogical methods of teaching staff. Professor Braquelaire sees these competences as transversal ones and argues that they should enhance general problem-solving skills of students. He states that, in his opinion, research skills of students should not be practiced within separate courses and that undergraduate research should be considered within the whole integrity of the study programs. However, concrete models of research training are still in the development phase.

Program-Level Initiatives for Supporting Undergraduate Research

Among the Université Côte d'Azur (UCA) initiatives, the Law School has introduced an optional workshop to start developing students' research skills. The workshop is available for students with the highest academic achievements. Apart from this, the research training for undergraduates is organized in the form of courses. The undergraduate courses aiming to introduce students to research combines theoretical and practical aspects of research but does not engage students in novel and authentic scientific data collection, analysis, and relevant or publishable work, such as in the context of Course-based Undergraduate Research Experiences (CUREs) being developed in the last few years in undergraduate courses (Hensel 2018), mostly located in the North American context. Alkaher and Dolan (2014) review different approaches for CUREs, engaging students in the "processes and practices of science" while conducting an empirical research where results advance scientific knowledge in a field. The context of CUREs engages students in real research activities, mostly in the field of the life sciences, especially in biology (Brownell and Kloser 2015). CUREs aim to support "students' development as scientists", including "learning about the nature and practice of science and building skills in doing science, including thinking like a scientist, reading and evaluating scientific literature, communicating about science, and collaborating with other scientists" (Dolan 2016, p. 6).

Course-Level Initiatives for Supporting Undergraduate Research

The most common model of undergraduate research in France is research embedded in the curriculum, through different types of courses aiming to introduce undergraduates to the research field. The courses are often located in the last year of the undergraduate curriculum to prepare those students who want to pursue their studies on a master's level. Other research introductory courses prepare undergraduate students for the undergraduate dissertation. These introductory research courses mostly focus on information search skills, the basis of scientific writing, and the correct use of references for properly citing the information sources. The examples provided below represent the course-level supporting undergraduate research at the University of Côte d'Azur. Undergraduate research is supported through courses within different disciplines such as musicology, chemistry, ethnology and anthropology, and law studies. We will introduce these initiatives in a more detailed way.

Law

At the Law School, undergraduate research training is carried out within excellence seminars. They are available to the 30 best students of the first year of bachelor studies. If they get good academic results, students are then supervised to the end of their undergraduate studies. The seminars last 16 hours per week and enable students to comprehensively develop their research projects.

Arts

Within the curriculum of musicology studies, there is a course on the introduction to the epistemology of this discipline. During the third year of their bachelor's program, each student should deliver a small-scale research project with the aim to prepare for the masters' studies.

Social Science and Humanities

The department of ethnology and anthropology proposes two courses on research in social sciences. A course on methodology and an internship that requires field research on a chosen topic, by using questionnaires, is also proposed within the second year of undergraduate-level studies.

Management and International Relations

A course dedicated to research is being prepared for the academic year 2019/2020, as well as funded 3-month research internships.

Sciences and Technologies

Since 2018, there is an undergraduate program, *Introduction to the research in chemistry*. Each semester, students attend courses providing research disciplinary training, but also general research training such as Scientific English. In the last semester of the third year, students should undertake a research internship within the university Fablab or another research lab. While the Université Côte d'Azur is a research-intensive site, we did not find undergraduate research examples in other areas of knowledge. Moreover, all these examples correspond to the way students are introduced to research, but they do not consider the engagement of undergraduate students in authentic research projects with existing labs. Some research labs of Université Côte d'Azur host undergraduate students. Within the Laboratoire d'Innovation et Numérique pour l'Education (LINE) we have offered undergraduate research fellowships to students from different backgrounds beginning in their freshman year. For example, this experience helped Fatma Ammar, an undergraduate student, to develop her confidence for considering the development of a PhD as a real possibility she can embrace.

Current Limits in Supporting Undergraduate Research in France

One important limit for supporting undergraduate research is the lack of recognition of undergraduate research as a key activity of academic positions. Promotion and tenure in France are mainly based on high-quality research outcomes such as research papers in peer-reviewed journals. Undergraduate research is not included in the official criteria for promotion or tenure evaluation. Faculty workload is mainly based on teaching hours, and when faculty oversee the introduction to research courses, these hours are considered in their workload. However, for other types of supervision of undergraduate research, there is no specific recognition of this teaching investment. Currently, we only identified a few financial incentives as, for instance, described in the Université Paris-Descartes call for projects (amount ranging from 10 to 15 K euros) for funding faculty-led activities that can initiate students to research activities.

Although undergraduate research competencies development is not a criterion for the promotion or tenure of faculty, we can appreciate the initiatives developed toward these objectives as personal or institutional activities that go beyond what is expected from undergraduate education and the faculty engaged in undergraduate programs. Furthermore, missions of research institutions like INRIA include the dissemination of research studies. It is dedicated to the public, ranging from children to adults and including also undergraduate students interested in knowing more about computer sciences research.

The lack of policy related to the way to engage and support undergraduate research activities and outcomes is also a limit, which could lead to bad practices. There is no policy to coauthor papers with undergraduate students or to copresent at regional, national, or international conferences. At Laboratoire d'Innovation et Numérique pour l'Education (LINE), we encourage the contribution of undergraduate students as coauthors when they contributed to research studies, but also teachers of primary and secondary level schools who have been engaged in collaborative research activities (Desgagné et al. 2001).

Opportunities to Develop Undergraduate Research in France

There is a need in France to develop a different representation of undergraduates to see them as possible actors of research and not only as students who should be introduced to research to raise their future interest in the field at the masters or doctorate level. A more participatory and collaborative orientation of research (Desgagné 1997) including citizen science (Oberhauser and LeBuhn 2012) could contribute to expanding the vision of current researchers, which should not be limited to graduate, doctorate, faculty, and research professionals. The general objective is twofold: to develop an awareness on the opportunities to support undergraduate research and to define students' contracts allowing the research labs to facilitate the engagement of undergraduate students to actively participate in the projects.

The organization of the undergraduate research activities within the new university curriculum could facilitate these experiences and develop a growing number of initiatives. In addition to formal activities, the organization of open doors at research labs could help to invite undergraduate students to better know the opportunities to be engaged in research. At the current moment, there are no undergraduate research celebration days in the universities. Actually, this type of celebration is organized by the doctoral schools but is focused on doctoral students having already started a PhD program. In terms of the modalities of engagement of undergraduate students, there are research labs and promoting the understanding of "contrats étudiants" (students' contract) among undergraduate students could help in facilitating research collaboration opportunities. Students' contract guarantees them a minimum wage by law. At the institutional level, at the university but also at the national level, we propose to raise the awareness of policymakers about the importance of undergraduate research to help them introduce policies to support these practices.

The assessment of the initiatives supporting undergraduate research could contribute to improving the existing initiatives in France. Currently, there is not a national-based assessment of undergraduate research in France, nor a framework for research competencies at the university level. Despite the nonexistence of assessment strategies, there is the study of Rinck (2016) in which she evaluates the capacity to develop writing research competencies among undergraduate students in their third year (L3) at the University of Paris 8.

The development of well-structured undergraduate research programs will require government leaders and HE decision-makers to develop an awareness of the existing international initiatives, such as CUREs (Brownell and Kloser 2015; Hensel 2018). The development of agreements and exchanges with other European countries within the Erasmus+ initiatives in terms of undergraduate research could also contribute to the development of undergraduate research in France by helping to transfer some of the successful international initiatives into the French HE institutions.

References

Alkaher, I., & Dolan, E. L. (2014). Integrating research into undergraduate courses. *Research Based Undergraduate Science Teaching*, 403–434.

Altman, J. D., Chiang, T.-M., Hamann, C. S., Makhluf, H., Peterson, V., & Orel, S. E. (2019). *Undergraduate research: A road map for meeting future national needs and competing in a world of change* (Council on undergraduate research). Washington, DC.

Bedenlier, S., Kondakci, Y., & Zawacki-Richter, O. (2018). Two decades of research into the internationalization of higher education: Major themes in the journal of studies in international education (1997–2016). *Journal of Studies in International Education, 22*(2), 108–135.

Brownell, S. E., & Kloser, M. J. (2015). Toward a conceptual framework for measuring the effectiveness of course-based undergraduate research experiences in undergraduate biology. *Studies in Higher Education, 40*(3), 525–544.

Cassone, L., Romero, M., Viéville, T., De Smet, C., & Ndiaye, M. (2019). Co-creativity, robotics and maker education proceedings. *ANR# CreaMaker Workshop*. Presented at the Nice, France. Retrieved from https://www.researchgate.net/publication/332383744_Co-creativity_robotics_and_maker_education

Chouinard, M.-C., Lavoie, M., Poitras, M.-È., Sasseville, M., de Jordy, L. C., & Girard, A. (2015). Utilisation d'un club de lecture comme méthode pédagogique pour favoriser l'appropriation des données probantes par des étudiantes au baccalauréat en sciences infirmières. *Recherche en Soins Infirmiers*, (1), 35–46.

Desgagné, S. (1997). Le concept de recherche collaborative: l'idée d'un rapprochement entre chercheurs universitaires et praticiens enseignants. *Revue Des Sciences de l'éducation, 23*(2), 371–393.

Desgagné, S., Bednarz, N., Lebuis, P., Poirier, L., & Couture, C. (2001). L'approche collaborative de recherche en éducation: un rapport nouveau à établir entre recherche et formation. *Revue Des Sciences de l'éducation, 27*(1), 33–64.

Dolan, E. L. (2016). *Course-based undergraduate research experiences: Current knowledge and future directions*. National Research Council Commissioned Paper, Washington, DC.

Garrett, R. (2019). Whatever happened to the promise of online learning? *International Higher Education*, (97), 2–4.

Hensel, N. H. (Ed.). (2018). *Course-based undergraduate research: Educational equity and high-impact practice* (1st ed.). Sterling: Stylus Publishing.

MESRI. (2018). *Organismes de recherche. Liste des principaux établissements publics de recherche*. Retrieved from Ministère de l'Enseignement supérieur et de la Recherche website: http://m.enseignementsup-recherche.gouv.fr/pid24575-cid49677/principaux-etablissements-publics-de-recherche-et-d-enseignement-superieur.html

Meunier, J.-M., & Zibetti, E. (2019). Penser, s'Etonner, Problématiser et Evaluer (PEPE): un parcours pluriannuel d'initiation à la recherche en licence à distance. *Education 4.1 Distances, Médiations Des Savoirs et Des Formations.*

Milyon Lab. (2017). La Laboratorie d'excellence Milyon. https://milyon.universite-lyon.fr/

Oberhauser, K., & LeBuhn, G. (2012). Insects and plants: Engaging undergraduates in authentic research through citizen science. *Frontiers in Ecology and the Environment, 10*(6), 318–320.

Open Access Government. (2019). *France: Investing in innovation, education and research.* Retrieved from Ministère de l'Enseignement supérieur et de la Recherche website: https://www.openaccessgovernment.org/investing-in-education/58921/

Regnaut, C. (2012). *Bilan de l'activité du CEVU. Université Paris-Est Marne-la-Vallée (UPEM). Février 2008 -Janvier 2012.* Retrieved from https://www.researchgate.net/profile/C_Regnaut/publication/272792225_Bilan_de_l'activite_du_CEVU_2008-2012_presente_au_nouveau_CEVU_2_avril_2012/links/54ee90ab0cf25238f93aa338/Bilan-de-lactivite-du-CEVU-2008-2012-presente-au-nouveau-CEVU-2-avril-2012.pdf

Rinck, F. (2016). Comment former des enseignants spécialistes de l'écriture et de son enseignement. *Pratiques*, (171–172). https://doi.org/10.4000/pratiques.3197.

Yang, R. (2018). *The third delight: Internationalization of higher education in China*. Routledge.

CHAPTER 12

Undergraduate Research in German Higher Education: Tradition, Policy, and Innovation

Wolfgang Deicke and Harald A. Mieg

Introduction

Germany occupies a somewhat paradoxical role with regard to undergraduate research. On the one hand, its tradition of involving students in research can be traced back to Wilhelm von Humboldt's sketches toward a modern research university in 1809/10. On the other hand, and for reasons also dating back to Humboldt, the concept of "undergraduate research" is a fairly recent addition to the German discourse of higher education learning and teaching. Part of Humboldt's far-reaching educational reforms in Prussia, later adopted throughout Germany, was to relegate "undergraduate studies"—the Baccalaureate—to the school system. With the *Abitur*, German grammar schools were supposed to produce

W. Deicke
bologna.lab, Humboldt-Universität zu Berlin, Berlin, Germany
e-mail: wolfgang.deicke@hu-berlin.de

H. A. Mieg (✉)
Georg-Simmel Center for Metropolitan Studies,
Humboldt-Universität zu Berlin, Berlin, Germany
e-mail: harald.mieg@hu-berlin.de

N. H. Hensel, P. Blessinger (eds.), *International Perspectives on Undergraduate Research*,
https://doi.org/10.1007/978-3-030-53559-9_12

students who were mature enough and educated in the classics, arts, languages, and sciences to apply themselves to independent study in more or less any subject on offer at university at the time. For most of their modern history, therefore, German universities offered only degrees at postgraduate level—the specialized diploma for the applied sciences (4–7 years), state examinations—the *Staatsexamen*—for teachers, lawyers, and theologians as well as research doctorates for the humanities. The Magister, a broader postgraduate qualification, was introduced in the 1950s to provide increasing numbers of students in the arts and humanities with an alternative to the state exam and an exit option below the doctorate. The Bachelor as first qualifying academic degree was reintroduced into German higher education with the European Bologna-Reform process from 1999 onward and has only really begun to affect and dominate higher education discourse with the phasing out of the old Magister and Diploma degrees between 2009 and 2013. In the first part of this chapter, we provide some historical context on the relationship between teaching and research in German Higher Education leading up to and including the early implementation of the Bologna reform process in Germany (1999–2009). In the second part, we examine recent trends and developments in the linkage between research and teaching in German Higher Education (2001–2014). In the third part, we then introduce some current models and examples for the implementation of research-based learning.

Teaching and Research in the German Higher Education Context

For the case of German higher education, it thus makes sense to distinguish between "undergraduate research" as a novel concept and the much richer and developed discourse around "research-based learning" or *Forschendes Lernen* (inquiry-based learning). Here, too, some historical context is in order, as the focus of the debate on the relationship between research and teaching has shifted considerably with each major wave of higher education reform and expansion. Four of these turns are worth noting in particular:

1. When Wilhelm von Humboldt developed his blueprint of the modern research university, it was partly to ward off the threat of universities being relegated to mere teaching institutions by the rival

academies of science. The *Universität zu Berlin* he helped to set up in 1809/10 was, with a total population of 52 professors to 256 students, also fairly well resourced to facilitate the ideal of a university as a small community of students and teachers, working together to advance the progress of science (Humboldt 1809/2010, p. 230).

2. A first major challenge to the Humboldtian ideal of a "unity of research and teaching" came with the expansion of the German university system at the turn of the twentieth century. Between 1891 and 1911, the number of students in higher education more than doubled from 35,200 to 73,600, reaching 119,400 by 1921 (Müller-Benedict 2016, p. 69). The foundation of a powerful and highly prestigious extramural network of research institutes in the *Kaiser-Wilhelm-Gesellschaft* in 1911 can in part be seen as a return to the division of labor between the universities (education) and the academies (research) that Humboldt was trying to fight in 1809/10. While the primary concern here was to secure the quality and quantity of research output to support Germany's economic and political ambitions at the time, the foundation of an extramural research organization also supported the idea that teaching could be something that gets in the way of scientific progress—and a career in science.
3. The origins of the current debate around "research-based learning" as a didactical concept can be traced back to the higher education reforms of the 1970s to the 1980s. Against the backdrop of another rapid expansion of the higher education system—in West Germany student figures rose from 427,200 (1970) to 1,408,700 (1987)—concerns emerged around the degree to which a modern mass higher education system could still guarantee the scientific nature of academic training. This time, the leading voice was the *Bundesassistentenkonferenz* (BAK, Congress of University Lecturers) who demanded that students—at university at the very least—should be "trained by scientists in a scientific discipline and for a scientific occupation" (BAK 1970, p. 9); that they should be able to independently choose a research topic and develop a research question; determine a research design; experience a research process; learn to act and think as members of the scientific community; reflect critically about the relationship between hypotheses, methods, and results; and be able to present their results (cf. Deicke et al. 2014, p. 27). While the BAK's demands for students' involvement in

ongoing (staff or third party) research projects were never systematically implemented, outside of the new 'reform' universities such as Bremen or Bielefeld, they spawned a number of new teaching and learning formats—most notably the "project seminar"—which were adopted more widely throughout German higher education and attempted to realize the demands of the BAK (Fichten et al. 1978; Wildt 1981).

While perhaps falling short of the ideals of "research-based learning" set out by the BAK, the Magister and Diploma degrees at university remained at least research-informed in their content, research-oriented in training, and also contained opportunities for independent student research (cf. Healey and Jenkins 2009). In the humanities and social sciences, these would take the form of extended pieces of research writing (the *Hausarbeit*) from early on in the degree program. Most subjects also required longer final year dissertation projects. It could be argued, however, that the link of the curricula to research remained dominated by staff interest rather than student learning and that—in times of rising student numbers—"opportunities for independent research", linked closely to summative assessment, could mean poorly supervised research in isolation.

4. A third major turn in the debate around the relationship between research and teaching came, as indicated above, with the Bologna Reform process and the reintroduction of the Bachelor into the German higher education system. Again, this coincided with a massive increase in student numbers from 1,773,956 students in 2000 to 2,844,978 by 2017 (Federal Office for Statistics 2019). The real challenge, however, was the concept of the three-year Bachelor as a first qualifying academic degree. While the rest of the world tried to shift their higher education systems from "teaching to learning" (Boyer Commission 1998), it is fair to say that—in the first round of introducing the new undergraduate degrees between 2000 and 2009—many German universities went the other way. In attempts to salvage as much content as possible from the 5–7-year-long Magister and Diploma degrees, they often went for delivery-based and highly condensed curricula with restricted student choice and a tendency toward overassessment. This, combined with the attempt to introduce student fees in a previously free public education sys-

tem in 2006–2007, predictably resulted in major protests by students (with considerable support from academic staff).

Research-Based Learning and Undergraduate Research After the Bologna-Reform

It is at this point that the BAK's concept of "research-based learning" was rediscovered and identified as a possible remedy for many (if not all) the ills of mass higher education. If its champions in the 1970s had been the numerically strong, but relatively powerless BAK, support this time came from the very top. In 2006, the *Wissenschaftsrat* (German Science Council), the most influential advisory body on German higher education policy, recommended that research-based learning should be a key component of all degree level training:

> *University training can qualify (graduates) for qualified work [...] where it aims to develop the ability to independently develop questions, to systematically engage with problems, methodically generate new insights and critically reflect on fundamental questions. This can be achieved by teaching that demonstrates and discusses the scientific process and actively involves students in this process. Research-based learning thus is essential to every (kind of) scientific program of studies.* (Wissenschaftsrat 2006, p. 64)

It is worth noting the slight shift, again, of emphasis here: Where the BAK was still very much concerned with the standard of scientific training for careers in science, the Science Council's concern shifted much more with the transfer of academic skills to a more generic, not necessarily scientific, (graduate)job market.

With regard to "undergraduate research", there is another important difference to the situation in the 1970s: This time, the German government provided funding not just to facilitate the rapid expanse of higher education, but also offered specific funding to enhance the quality of teaching. Following the Science Council's advice, programs to support and implement "research-based learning" were very prominent amongst the measures funded in the two rounds of the *Qualitätspakt Lehre* (Quality Pact for Teaching, 2010–2016, 2016–2020). Perhaps the most notable innovation that resulted from this was the university-wide, cross-faculty *Research Opportunities Programs* piloted in a number of Quality Pact projects. The Quality Pact for Teaching has served as a catalyst for promoting and advancing research-based learning and specifically research-based learning opportunities aimed at undergraduate students.

Around these programs, an active network of institutions, QPT-projects and individuals has formed. To date, the network generated successful bids for two funded research projects into how research-based-learning works in different formats and disciplines ("ForschenLernen", 2014–2018, cf. Gess et al. 2017; Ouelette et al. 2017; Wessels et al. 2018; Gess et al. 2019; Mieg 2019a) and how it can be used in the crucial first year of study to enhance the student experience and strengthen retention ("Forschendes Lernen in der Studieneingangsphase (FideS)"). In 2016, members of the network established an interdisciplinary national conference for student research (Twitter hashtag #stufo20xx) that is now entering its fourth year. Carl-von-Ossietzky Universität, Oldenburg, was also selected to host the Second World Conference for Undergraduate Research in May 2019. Comprehensive overviews of the range of institutions and projects involved in the network and the type of issues they wrestle with can be found in the volumes edited by Mieg (2019b), Lehmann and Mieg (2018), and Mieg and Lehmann (2017).

While there is, to date, no national organization focused on the promotion of research-based learning, a large number of individuals from the RBL-community are organized in a standing group within the *Deutsche Gesellschaft für Hochschuldidaktik* (DGHD, German Association for the Professionalization of Teaching in Higher Education), the so-called AGFL (Working Group Research-based Learning). The group was initiated by Harald Mieg and Ludwig Huber in 2014 and—with 130+ active members from 57 institutions—is today one of the largest and most active working groups within the DGHD. The group holds regular meetings and workshops 2–4 times a year, with smaller subgroups collaborating digitally on selected themes. It has established a working papers series and is working on a collection of materials of use for preparing researchers, teaching staff, and students for research-based learning. Since 2013, the DGHD's annual conference has included a strong stream of contributions (empirical papers, SoTL cases studies, workshops) on research-based learning. Member institutions have hosted several research conferences around student and undergraduate research, the latest being the focus URE Conference at Universität Hohenheim in June 2019.

Current Trends and Debates on Research-Based Learning

The focus of the theoretical/conceptual debate in Germany between 2012 and 2016 was very much on matching the established German discourse around "Forschendes Lernen" (inquiry-based learning) to the new contexts of Bachelor and Master degrees and mapping it against the host of concepts emerging from the Anglo-American debates around the research–teaching nexus and other formats of active, student-centered learning and teaching. The concept of "undergraduate research" and the "undergraduate research experience" have so far played a subordinate role in these discussions.

While there are a number of competing German approaches to research-based education, two main paradigms can be identified: On the one hand, a competencies- and outcomes-based approach rooted in empirical educational psychology that tends to prefer "research-oriented teaching" (e.g. Reiber 2007; Schneider and Wildt 2009), and, on the other hand, a more pedagogically informed and process-oriented approach that tends to refer to "research-based learning" (Huber 2009, 2014; Tremp and Hildbrand 2012; Sonntag et al. 2016). While these differences in approach can lead to differences in the implementation of student research opportunity programs (e.g. opportunities to acquire or deepen specific sets of methodological skills and theoretical competencies vs. opportunities to develop or participate actively in actual research projects), there is a pragmatic consensus within the UGR/RBL-community that research and teaching can and must be (more) actively linked at all levels of under- and postgraduate degree programs.

There is also considerable convergence between the German and the international debates surrounding student and undergraduate research. Building on the works of Beckman and Hensel (2009), Brew (2006, 2013), Healey (2005), and Healey and Jenkins (2009), the German debate has produced several models for mapping the research–teaching nexus (Tremp and Hildbrand 2012; Rueß et al. 2013) and planning for research-based learning (Lübcke et al. 2017).

Despite the earlier criticisms of increasingly condensed undergraduate curricula, it is fair to say that German higher education has maintained very close links to research. While there are differences between academic subjects, this most certainly applies to all universities, which consider the link to research and—at the very least—research-informed and -oriented curricula to be the very feature that distinguishes them from the more applied universities and colleges of higher education. Here, and with many of the technical universities, the answer depends on whether, or to what extent, we count creative (Arts, Design) and constructive processes (Engineering) as research. In general, data from the national student surveys 2001–2013 (Ramm et al. 2014, pp. 38–9, cf. Table 12.1) appears to suggest two things: One that the link between research and teaching is becoming stronger across all types of institutions; and two, that the gap between university and more applied subjects appears to be shrinking and German higher education as a whole appears to be back on the track to shift from "teaching to learning".

If differences between types of higher education institutions appear to be diminishing, data from the national student survey shows that differences in research training (research-oriented teaching and learning opportunities) and opportunities for undergraduate research between the disciplines remain. The most recent data available comes from Bargel and Multrus' (2012) analysis of national student survey data from 2009–2011. According to this, links to research are considered most important in the natural sciences at university and least pronounced in law and economics at both levels, with the (mostly administrative) law-related degrees in the

Table 12.1 Students' perceptions of links between research and teaching 2001–2013 (percentages, by type of institution)

	Strong links	*Links exist*
Universities		
2013	31	44
2010	24	47
2007	22	47
2004	19	46
2001	18	44
Applied universities/Colleges of higher education		
2013	22	45
2010	15	43
2007	14	44
2004	10	42
2001	6	36

Data source: Studierendensurvey 1983–2013, AG Hochschulforschung Universität Konstanz

Adapted from: Ramm, Multrus, Bargel and Schmidt (2014, p. 39)

Applied Universities and Colleges of Higher Education taking the bottom spot. There is, at least in economics, a difference between more research-oriented pure economics (*Volkswirtschaftslehre*) and more applied business studies (*Betriebswirtschaftslehre*) degrees and some interesting attempts to remedy the lack of research opportunities in the latter (cf. Müller-Christ 2019). What was interesting to note here is that the links between research and teaching appear less pronounced in degrees in medicine at university than they are in the newly academicized nursing and health science degrees in the applied institutions. It seems that these professions have benefitted most from the initial wave of the Bologna reforms and the switch from vocational to academic training (Table 12.2), with university-based medical training programs—for example, at Charité Berlin or the Ruhr-Universität Bochum—now following up with new model degrees (cf. Schäfer 2019).

Table 12.2 Students' perception of the importance of links between research and teaching in Bachelor degrees at University and in Applied Colleges of Higher Education (percentages, by disciplines)

The link between research and teaching is important	*Cultural science*	*Social science*	*Law*	*Economics*	*Health sciences*	*Natural sciences*	*Agri-cultural science*	*Engi-neering sciences*
Universities								
In class	56	62	39	45	58	**70**	68	65
Specialized classes	48	51	34	40	49	**58**	57	53
Own participation	51	55	31	37	44	**69**	68	60
Applied universities								
In class	47	58	30	41	**68**	54	63	55
Specialized classes	43	51	33	37	**63**	49	55	49
Own participation	42	49	29	34	57	54	**64**	56

Data source: Studienqualitätsmonitor 2009–2011, HIS Hannover & Universität Konstanz

Adapted from: Bargel and Multrus (2012, p. 14), **top ranks** highlighted in, bold

Research-Based Learning in Practice: Examples and Models

While the data generally reflects our own curricular analysis at Humboldt-Universität zu Berlin, they obscure another important dimension of undergraduate research: the degrees of freedom that students are granted in developing and carrying out independent research projects at the undergraduate level. While opportunities to participate in research may be most plentiful in the natural sciences, it is less likely for undergraduate STEM-students to be allowed to develop their own research questions, choose a research design or method than it is in the social sciences. Undergraduate students in the natural sciences will more likely be assigned a problem, question, or experiment to conduct, analyze, and report on for their final year project (cf. Ruf et al. 2019, pp. 200–1).

This is also reflected in the predominant models of research training. Insofar as Humboldt-Universität can serve as a typical example for a research-active university, laboratory training and the apprentice model are the most common modes of training in the natural sciences. Here, undergraduate students can apply for paid positions as student assistants in third-party-funded research projects. They are usually assigned to particular tasks in the project and can write their undergraduate dissertation or a research paper under the guidance of a junior researcher and/or the supervision of a principal investigator.

Research projects embedded in the curriculum tend to be more common in the social sciences and humanities. These are often structured as social, rather than individual projects. A key example from Humboldt-Universität would be the compulsory year-long final year fieldwork project, in which groups of students assign themselves to one of a number of set fieldwork projects, begin to explore the field, and gain practical experience of a range of ethnological research methods under the supervision of an experienced academic and then begin to develop their own questions and approaches to a particular aspect in that field. Ideally, the joint project (and the data produced through it) and/or the methodological skills acquired are then used or followed up in the students' individual BA-dissertations.

With the notable exceptions of Leuphana University, Lüneburg (cf. Lang and Wiek 2013) and the University of Bremen (cf. Huber et al. 2013), community-based research is still relatively rare in German universities. Projects like the several Quality-Pact funded law clinics at

Humboldt-Universität (Human Rights and Citizenship Law, Consumer Rights, Internet Law, and the student-initiated and -led Refugee Law Clinic) indicate, however, that this is changing and that subjects perceived as not having strong active links between research and the curriculum are particularly open to exploring these "novel" ways of linking professional training with research activities. Community-based research and service-learning models tend to be more common in the social sciences and health sciences programs in Universities and HE Colleges of Applied Science (Schmidt-Wenzel and Rubel 2019). Among the Quality-Pact funded Student Research Opportunities Schemes, there are several creative attempts to bring the research from Germany's extramural research institutes back into teaching. A good example for this is the Q-Teams created by Humboldt-Universität's bologna.lab. In this format, junior researchers in one of Berlin's 70+ extramural research institutes can apply for funding for a small student research team to work on a particular aspect of one of the institute's ongoing research projects on the understanding that they are not considered "cheap additional labor", but trained and mentored as future researchers. In addition to a paid temporary teaching contract, the junior researchers receive didactical training and support in planning the project and acquire additional skills both in teaching as well as in leading a (student) research team, which are useful for their next steps in postdoctoral careers in teaching and/or research (cf. Deicke et al. 2014).

An additional model, far more common in universities of applied science, is the dual career degree, which combines vocational/professional training in the workplace with degree level studies (the combination of vocational schooling and workplace training has always been a general strength of the German education system). The leading German exponent of this is the *Duale Hochschule Baden-Württemberg* (2019) and a number of private Universities such as *Steinbeis Hochschule Berlin* offer dual-career degrees. However, while German research universities would have turned their nose at the mere idea of "application" a decade ago, there are signs that the (faux) distinction between research and practice is softening here, too. At Humboldt-Universität, the introduction of research-based project work in undergraduate teacher training degrees now goes hand-in-glove with the introduction of a dual-career postgraduate teaching qualification (Master of Education) for graduates and applicants with advanced vocational/professional training and/or substantive work experience. Similarly, many higher education institutions now support their researchers and students in taking the products, innovations, and designs from their research to the market. At Humboldt-Universität, this is facilitated through Humboldt Innovation, a private company founded in 2005 and owned 100% by Humboldt-Universität (a public institution). Humboldt Innovation provides grants, legal and financial advice and office space to students, graduates, and researchers from Humboldt-Universität who have an idea or product ready for taking to the market. While not primarily aimed at undergraduate students and more likely to be used by graduate and researchers between master's level and postdoc, it has supported a number of start-ups by undergraduate and graduate students from different faculties and successfully involves undergraduates in its information sessions and social events (Humboldt Innovation 2019).

The federal government's strategy and support for undergraduate research is rather more difficult to assess. While there is a lot of public funding for research and development, the bulk of it (still) goes to independent research societies such as the *Max Planck Gesellschaft*, the *Leibniz Gesellschaft*, the *Helmholtz Gesellschaft*, and the *Fraunhofer Institutes*, and therefore straight past the bulk of the undergraduate students. University professors can apply for funding to the *Deutsche Forschungsgemeinschaft* (DFG, *German Research Society*), but usually use the funding they receive for a teaching "buy out". In the first two rounds of the German high-profile *Excellence Initiative*, universities could include graduate schools and structured PhD programs in their proposal but were not allowed to use any of their research funds to support undergraduate initiatives. While this rule was relaxed for the most recent round of applications and research-based teaching was mentioned in both the calls for research clusters as well as bids for the status of *Excellence University*, the DFG's and German Science Council's Call for Submissions in 2016 tellingly still described teaching as "(one) of the other ancillary functions of universities" (DFG 2016, p. 3). Consequently, the Berlin University Alliance's successful bid for funding includes a research-based component modeled on Humboldt-Universität's HU-Q program and aimed predominantly at undergraduate students, but with an annual budget of just €1.5 mn out of €25 mn in total.

For Universities of Applied Sciences, there have been several federal funding lines for research infrastructure and the development of research profiles since 1992, currently grouped under the umbrella of *Forschung an Fachhochschulen*(research in universities of applied science). The increase in funding under this umbrella—from €10.5 mn in 2005 to €48 mn in 2016 (BMBF 2016, p. 6)—may in part explain the changes in the perceived linkage between research and teaching in the applied universities and colleges of higher education. While the applied universities have been very successful in attracting third-party funding, their figures pale into insignificance compared to the annual budget of the *Deutsche Forschungsgemeinschaft*, which stood at €3.4 bn per annum in 2018 (DFG 2018, p. 2). Tellingly, too, a recent program evaluation of the funding line *Forschung an Fachhochschulen* (Geyer et al. 2016) covers many dimensions of research in applied universities, but completely omits teaching—suggesting that the federal and state governments as well as many institutional leaders—are missing out on something important by aspiring to become more like universities, who in turn aspire to become more like the independent research institutes.

Conclusion

Despite these misgivings, we would conclude that research-based learning and—to a lesser extent—undergraduate research have taken a strong hold in German higher education discourse and policy. While there are still massive disparities in esteem and importance between teaching and research, research-based learning has the potential—as its (marginal) inclusion in the *Excellence Strategy* and funding through the Quality Pact have demonstrated—to somewhat redress this gap.

For the reasons outlined above, there are, at present, very few days or events aimed at celebrating undergraduate or student research for its own sake. There are student-organized and university-funded research conferences (such as the Sustainability Conference in Berlin), but these tend to be subject-focused rather than aimed at creating visibility for student research as an end in itself. The closest equivalent we have at present is the annual conference for student research, established in 2016. This is a multidisciplinary, national conference aimed at providing a platform for student research and has to date been hosted by Carl-von-Ossietzky Universität, Oldenburg; Humboldt-Universität zu Berlin (2017); Ruhr-Universität Bochum (2018).

The main challenge at present arises from the temporary nature of the Quality Pact for Teaching and chronically underfunded Higher Education Institutions. The per capita funding HEIs receive per student has not kept up with the rapid increase in student numbers since 2000. This means that, when the current Quality Pact funding period ends in December 2020, many universities will be faced with tough choices regarding the allocation of resources. On the whole, it seems fair to conclude that German higher education has overcome the temporary setbacks and problems linked to the implementation of the Bologna reform process. As always in politics or education, change takes time and will only be complete in this instance once the academic CV of the next president of the German rectors' conference reads BA, MA, PhD habil.

References

Bargel, T., & Multrus, F. (2012). *Das Bachelorstudium im Spiegel des Studienqualitätsmonitors: Entwicklungen der Studienbedingungen und Studienqualität 2009 bis 2011.* Hannover/Konstanz, Germany. Hochschulinformations-System & AG Hochschulforschung.

Beckman, M., & Hensel, N. (2009). Making explicit the implicit: Defining undergraduate research. *CUR Quarterly, 4*, 40–44.

Boyer Commission. (1998). Reinventing Undergraduate Education: A Blueprint for America's Research Universities. Boyer Commission on Educating Undergraduates in the Research University. Stoney Brook. New York.

Brew, A. (2006). *Research and teaching: Beyond the divide.* London: Palgrave Macmillan.

Brew, A. (2013). Understanding the scope of undergraduate research: A framework for curricular and pedagogical decision-making. *Higher Education, 66*(5), 603–618.

Bundesassistentenkonferenz (BAK). (1970). *Forschendes Lernen – Wissenschaftliches Prüfen. Ergebnisse der Arbeit des Ausschusses für Hochschuldidaktik, Bonn* (Nachdruck 2009). Bielefeld: Universitäts Verlag Webler.

Bundesministerium für Bildung und Forschung (BMBF). (2016). *Forschung an Fachhochschulen.* Bonn/Berlin: Federal Ministry of Education and Research.

Deicke, W., Gess, C., & Rueß, J. (2014). Increasing students' research interest through research-based learning at Humboldt-University. *CUR Quarterly, 35*(1), 27–33.

Deutsche Forschungsgemeinschaft (DFG). (2016). Exzellenzstrategie des Bundes und der Länder Ausschreibung für die Förderlinie Exzellenzuniversitäten. (Förderphase 1. November 2019 bis 31. Oktober 2026). *Call for Applications,* 28. September 2016.

Deutsche Forschungsgemeinschaft (DFG). (2018). *Facts and figures 2018.* Bonn: DFG.

Duale Hochschule Baden-Württemberg. (2019). *Homepage.* Available at http://www.dhbw.de/english/dhbw/about-us.html. Last accessed 07 Aug 2019.

Federal Office for Statistics. (2019). Statistics of students – (Table 21311–0001 in English). https://www-genesis.destatis.de/genesis/online/data;sid=DA

7B78724113716E96C51F2D8E3D20C6.GO_1_3?operation=abruftabelleBearbeiten&levelindex=2&levelid=1563123054067&auswahloperation=abruftabelleAuspraegungAuswaehlen&auswahlverzeichnis=ordnungsstruktur&auswahlziel=werteabruf&selectionname=21311-0001&auswahltext=&werteabruf=start on.html

Fichten, W., Jaeckel, K., & Stinshoff, R. (Eds.). (1978). *Projektstudium und Praxisbezug. Reformmodelle der Lehrer- und Juristenausbildung*. Frankfurt/Main: Campus Verlag.

Gess, C., Wessels, I., & Blömeke, S. (2017). Domain-specificity of research competencies in the social sciences: Evidence from differential item functioning. *Journal for Educational Research Online/Journal Für Bildungsforschung Online, 9*(2), 11–36.

Gess, C., Deicke, W., & Wessels, I. (2019). Competence development through inquiry-based learning. In H. A. Mieg (Ed.), *Undergraduate research. The German multidisciplinary experience*. Cham: Springer Open.

Geyer, A., Berger, F., & Dudenbostel, T. (2016). *Empirische Begleitung des Programms „Forschung an Fachhochschulen"*. Abschlussbericht. Vienna: Technopolis group.

Healey, M. (2005). Linking research and teaching: Exploring disciplinary spaces and the role of inquiry-based learning. In R. Barnett (Ed.), *Reshaping the university: New relationships between research, scholarship and teaching* (pp. 67–78). Berkshire: McGraw Hill/Open University Press.

Healey, M., & Jenkins, A. (2009). *Developing undergraduate research and inquiry*. York: HE Academy. Available online at https://www.heacademy.ac.uk/assets/York/documents/resources/publications/DevelopingUndergraduate_Final.pdf. Last accessed 15 Apr 2019.

Huber, L. (2009). Warum Forschendes Lernen nötig und möglich ist. In L. Huber, J. Hellmer, & F. Schneider (Eds.), *Forschendes Lernen im Studium, Aktuelle Konzepte und Erfahrungen* (pp. 9–35). Bielefeld: UVW.

Huber, L. (2014). Forschungsbasiertes, Forschungsorientiertes, Forschendes Lernen: Alles dasselbe? *Das Hochschulwesen, 62*(1+2), 22–29.

Huber, L., Kröger, M., & Schelhowe, H. (Eds.). (2013). *Forschendes Lernen als Profilmerkmal einer Universität. Beispiele aus der Universität Bremen*. Bielefeld: Universitäts Verlag Webler.

Humboldt Innovation. (2019). *Research creating reality*. Homepage. Available at https://www.humboldt-innovation.de/en/aboutus.html. Date last accessed 07 Aug 2019.

Humboldt, W. von. (1809/2010). *Über die innere und äußere Verfassung der Universität*. Berlin: Humboldt-Universität zu Berlin.

Lang, D., & Wiek, A. (2013). The role of universities in fostering urban and regional sustainability. In H. A. Mieg & K. Töpfer (Eds.), *Institutional and social innovation for sustainable urban development* (pp. 393–411). London: Earthscan.

Lehmann, J., & Mieg, H. A. (Eds.). (2018). *Forschendes Lernen: Ein Praxisbuch*. Potsdam: Verlag der Fachhochschule Potsdam.

Lübcke, E., Reinmann, G., & Heudorfer, A. (2017). Entwicklung eines Instruments zur Analyse Forschenden Lernens. *Zeitschrift für Hochschulentwicklung, 12*(2), 191–216.

Mieg, H. A. (2019a). Forms ofresearchwithinstrategiesforimplementingundergraduateresearch. *ZFHE, 14*(1), 79–94.

Mieg, H. A. (Ed.). (2019b). *Inquiry-based learning—Undergraduate research: The German multidisciplinary experience*. Cham: Springer Open. (open access, translation of Mieg& Lehmann, 2017).

Mieg, H. A., & Lehmann, J. (Eds.). (2017). *Forschendes Lernen: Wie die Lehre in Universität und Fachhochschule erneuert werden kann*. Frankfurt/Main: Campus.

Müller-Benedict, V. (2016, March 14). *Education and science*. www.deutschland-in-daten.de. http://www.deutschland-in-daten.de/en/migration. Last accessed 14 July 2019.

Müller-Christ, G. (2019). Inquiry-based learning in business administration. In H. A. Mieg (Ed.), *Inquiry-based learning – Undergraduate research. The German multidisciplinary experience* (pp. 261–270). Cham: Springer Open.

Ouelette, D. J., Zottmann, J., Bolzer, M., Fischer, F., & Fischer, M. R. (2017). Epistemological beliefs and scientific reasoning and argumentation. In H. Laitko, H. A. Mieg, & H. Parthey (Eds.), *Forschendes Lernen: Wissenschaftsforschung Jahrbuch* (pp. 136–151).

Ramm, M., Multrus, F., Bargel, T., & Schmidt, M. (2014). *Studiensituationen und studentische Orientierungen. 12. Studierendensurvey an Universitäten und Fachhochschulen*. Bonn/Berlin: Federal Ministry for Education and Research.

Reiber, K. (Ed.). (2007). *Forschendes Lernen als hochschuldidaktisches Prinzip – Grundlegung und Beispiele*. Tübinger Beiträge zur Hochschuldidaktik (online available.

Rueß, J., Gess, C., & Deicke, W. (2013). Schärfung des Konzepts Forschenden Lernens im Kontext forschungsorientierter Lehre. Presentation at the Conference *Forschendes Lernen: Forum für gute Lehre*, 2 September 2013, FH Potsdam.

Ruf, A., Ahrenholz, I., & Matthé, S. (2019). Inquiry-Based Learning in the Natural Sciences. In H. Mieg (Ed.), *Inquiry-based Learning – Undergraduate research. The German multidisciplinary experience* (pp. 193–204). Cham: Springer Open.

Schäfer, T. (2019). Iquiry-based learning in medicine. In H. A. Mieg (Ed.), *Inquiry-based learning – Undergraduate research. The German multidisciplinary experience* (pp. 161–170). Cham: Springer Open.

Schmidt-Wenzel, A., & Rubel, K. (2019). Inquiry-based learning in social work. In H. A. Mieg (Ed.), *Inquiry-based learning – Undergraduate research. The German multidisciplinary experience* (pp. 139–147). Cham: Springer Open.

Schneider, R., & Wildt, J. (2009). Forschendes Lernen und Kompetenzentwicklung. In L. Huber, J. Hellmer, & F. Schneider (Eds.), *Forschendes Lernen im Studium* (pp. 53–68). Bielefeld: Universitäts Verlag Webler.

Sonntag, M., Rueß, J., Ebert, C., Friederici, K., & Deicke, W. (2016). *Forschendes Lernen im Seminar. Ein Leitfaden für Lehrende.* Berlin: bologna.lab der Humboldt-Universität zu Berlin.

Tremp, P., & Hildbrand, T. (2012). Forschungsorientiertes Studium – universitäre Lehre: Das »Zürcher Framework« zur Verknüpfung von Lehre und Forschung. In T. Brinker& P. Tremp (Eds.), *Einführung in die Studiengangentwicklung* (pp. 101–116). Bielefeld: Bertelsmann.

Wessels, I., Rueß, J., Jenßen, L., Gess, C., & Deicke, W. (2018). Beyond cognition: Experts' views on affective-motivational research dispositions in the social sciences. *Frontiers in Psychology, 9*, 1300.

Wildt, J. (1981). *Hochschuldidaktik und staatliche Studienreform. Zur Transformation des Projektstudiums im Spannungsfeld einer Studienreform von „oben" und „unten"* (Materialien und Berichte, Vol. 13). Bielefeld: Interdisziplinäres Zentrum für Hochschuldidaktik der Universität Bielefeld.

Wissenschaftsrat. (2006). *Empfehlungen zur künftigen Rolle der Universitäten im Wissenschaftssystem.* Berlin.

CHAPTER 13

A Closer Look at Undergraduate Research in the United Arab Emirates

Jase Moussa-Inaty

A Drive Towards Undergraduate Research

The UAE has been at the heart of educational reform for quite some time and still it continues to enable and instill a culture of undergraduate research and innovation among new generations. For a country of only forty-seven years, many developments have come to fruition including transforming the UAE into a knowledge-based economy (Ahmed and Abdalla Alfaki 2013) and developments in the area of undergraduate research. It is argued that a focus on undergraduate research can entirely serve the UAE's vision to further improve the UAE education system. To help drive this reform, Emirati students are often encouraged to engage in research projects that serve the UAE's mission and vision. The consequence of an increased number of student-led research projects requires an availability of qualified professionals at a time of educational reform and change. Although research intensive universities in the UAE may be lacking (Ryan and Daly 2019) and despite several challenges related to the

J. Moussa-Inaty (✉)
Psychology, Zayed University, Dubai, United Arab Emirates
e-mail: jase.inaty@zu.ac.ae

N. H. Hensel, P. Blessinger (eds.), *International Perspectives on Undergraduate Research*,
https://doi.org/10.1007/978-3-030-53559-9_13

integration of innovation (Halaweh 2019), Emiratis are positively and persistently doing their best to lead and at times keep up with the rapidly changing scientific innovations both nationally and internationally.

UAE higher education programs and federal institutions also strive to be a part of an international trend of best practices in undergraduate research. The process of routing this goal for improvement and international recognition has urged higher education institutions in the UAE to continue to evolve where new campuses are continuously being developed to serve this need. With this in mind, there has also been a greater focus on research at both the graduate and undergraduate levels.

From a government standpoint, the UAE's Ministry of Education (MOE) governs the education sector and has laid out a strategic plan that includes developing an innovative educational system with research as one of eight of its main goals. The eight goals to be achieved by the year 2021 as stated on the Ministry's homepage (see https://www.moe.gov.ae) are:

1. Ensure inclusive quality education including preschool education
2. Achieve excellent leadership and educational efficiency
3. Ensure quality, efficiency and good governance of educational and institutional performance, including the delivery of teaching
4. Ensure a safe, conducive and challenging learning environment
5. Attract and prepare students to enroll in higher education internally and externally, in the light of labor market needs
6. Strengthen the capacity for scientific research and innovation in accordance with the quality, efficiency and transparency standards
7. Provision of quality, efficient and transparent administrative services, in accordance with the related standards
8. Establish a culture of innovation in an institutional working environment

Undergraduate Research Support

While the importance of undergraduate research is not explicitly indicated in the Ministry of Education's mission, vision, and strategic plan, the Ministry has shown support towards various national higher education institution initiatives that encourage undergraduate research. When reviewing undergraduate research in the UAE, quite a few higher education institutions stand out, such as Abu Dhabi University, Zayed University,

United Arab Emirates University, New York University (Abu Dhabi), and the American University of Ras Al Khaimah. The following sections will briefly discuss some of the undergraduate research highlights in these institutions.

Undergraduate Research at Abu Dhabi University

One of the country's largest initiatives is led by Abu Dhabi University—a federal institution based in Abu Dhabi with additional three campuses across Abu Dhabi, Dubai, and Al Dhafra. Its Office of Research and Sponsored Programs has been organizing an Undergraduate Research Competition (URC) for seven years in a row as part of its mission in order to *expand its research activities and build a reputation as a research-oriented university recognized nationally, regionally, and internationally*, as noted on the university website. The competition is held at Abu Dhabi University and is under the patronage of H. E. Hussain Ibrahim Al Hammadi, the Minister of Education. The goal of the competition is to encourage "universities to promote scientific research among undergraduate students and to make it an integral part of university education, given the significance of scientific research in advancing the country to the top ranks. It also aims to instill life-long learning, and to enhance career opportunities for young people as well as foster entrepreneurial mindsets and attitudes in education, research, and innovation". Although the competition is held at this university, it is open to all undergraduate students across UAE-based universities whether government-sponsored or private. This year and for the first time since the competition was launched, undergraduate students from Bahrain, Kuwait, Oman, Qatar, and Saudi Arabia were invited to compete.

Furthermore, there has been a significant increase in paper submissions ever since its inauguration with over two hundred and fifty papers presented that include individual or team-based projects. These projects are research-based in that some form of data collection and analysis, or critical review is conducted. The competition highlights basic and applied original research in six fields that include engineering, business administration, arts and social sciences, education and law, natural and health sciences, and technology and innovation. The review process after submission is rigorous and includes three rounds of evaluation by an external committee whose members also serve as judges. Winners of the competition are invited to present their projects at an organized symposium with a large number of invited guests and VIP members from the community and the

MOE in attendance. This year and for the URC 2019, an attractive amount of over 100, 000 UAE dirhams was distributed amongst the winners, with certificates of recognition and participation also given out to the participating undergraduate students during an award ceremony. This year, the competition's community strategic partner known as ADNOC (Abu Dhabi National Oil and Gas Company) sponsored and helped organize the awards ceremony. Over two hundred teams participated and

Table 13.1 List of URC 2019 winning universities

Number	*Country*	*University name*	*Number of winners*
1	The United Arab Emirates	University of Sharja	6
2	The United Arab Emirates	Abu Dhabi University	6
3	The United Arab Emirates	Khalifa University	3
4	The United Arab Emirates	American University of Sharjah	2
5	The United Arab Emirates	United Arab Emirates University	2
6	Saudi Arabia	King Saud University	2
7	The United Arab Emirates	Al Ain University of Science and Technology	2
8	The United Arab Emirates	University of Wollongong	2
9	The United Arab Emirates	New York University	1
10	Oman	Sultan Qaboos University	1
11	Oman	Dhofar University	1
12	The United Arab Emirates	Gulf Medical University	1
13	The United Arab Emirates	Birla institute of Technology and Science	1
14	The United Arab Emirates	Rochester Institute of Technology	1
15	The United Arab Emirates	Zayed University	1
16	The United Arab Emirates	Canadian University	1
17	The United Arab Emirates	Middlesex University	1

thirty-four winning teams were identified. Table 13.1 shows a list of winning universities at the URC 2019. A total of seventeen universities in the UAE, Saudi Arabia, and Oman comprised the winning teams (Sarrildin 2019).

Undergraduate Research at Zayed University

Another university supported by the Ministry of Education is Zayed University—one of three federal institutions in the UAE. Originally established over two decades ago to educate female Emirati students only, today Zayed University accepts both male and female Emirati and international students spread across two of its campuses located in Abu Dhabi and Dubai. Though research engagement is a common practice (Khelifa et al. 2004), there has been a recent push toward more research commitment at the undergraduate level. With five varying colleges and over fifteen programs to choose from, there is much room for research engagement at the undergraduate level.

Typically, students can join one of seven colleges: Colleges of Arts and Creative Enterprises; Business; Communication and Media Sciences; Education; Humanities and Social Sciences; Natural and Health Sciences; and Technological Innovation. University College as a college per se does not offer a degree; rather it offers a range of core courses common to all other colleges, meaning that all students enrolled in a major at Zayed University have to take all the courses offered by University College. One such course is GEN185 Methods of Scientific Research and Development that aims to teach research methods and allows students an opportunity to engage in undergraduate research. For example, as part of the GEN185 course final project enrolled students are required to conduct a small-scale research investigation and present their findings in a research paper and poster. Students may choose from wide range of topics that vary from year to year (see Table 13.2).

Elements of the paper should include a clear hypothesis and attempt to answer one or more research question. Furthermore, it should include the following sections: introduction, literature review, methodology, data collection, results, discussions, and a conclusion. The paper should also follow APA style and make reference to current and groundbreaking research-based sources. Students receive specific guidelines and one-on-one support if needed in order to undertake this type of research. Zayed University library and library commons also offer support in the area of

Table 13.2 GEN 185 Methods of Scientific Research and Development final project sample research topics

Topic	*Topic description*
Sleep habits	Sleep habits can be affected by many things that we do during our day. As a group, investigate the various factors that can lead to poor sleeping habits.
Washing soaps/ detergents	The foaming ability of a detergent /washing soap depends on several physical/chemical factors. Investigate the factors that affect the foaming capacity of detergents/washing soaps (e.g. the effect of hardness of water (dissolved salts) on foaming capacity, effect of water temperature on foaming capacity, effect of pH of water on foaming capacity, etc.)
Daycare	Many parents like daycare centers because they offer a formal, structured environment. Other parents might question how the behavior of a child who goes to daycare compares to a child that stays at home with parents. Depending on various circumstances, a child who attends daycare may be better prepared than a child who stays home with their parents. While an ordinary daycare center may watch over your child with safety in mind until pickup, an extraordinary daycare center may dedicate itself to helping prepare your child for the future. As a group, investigate the pros and cons of daycare centers.
Media and study	Many people like to have something playing when they study, whether music or video. Others prefer to turn off all distractions, including their phone. As a group, investigate the effect of different media on people's ability to memorize something.
Contemporary ideas of beauty among the local population in the UAE	What people consider beautiful is both subjective and influenced by culture and media, as well as age, and social and ethnic affiliation. Explore ideas of beauty in three different categories: (a) according to millennials (18–35 year olds), (b) according to the older generations, (c) according to the media. Each student in the group can address one of those categories, interviewing people and recording their views, looking at media depictions of "beautiful" people but not necessarily centered on people only. Things like fashion and places can also be examined to come to a conclusion about what people in the UAE think of "beauty" today.
Student transition	Transitioning from high school to university can be filled with a range of emotions and experiences. For example, some students will find the transition to be positive while others might find it to be negative or stressful. Explore the factors that affect student transitions from high school to university or investigate students' perceptions of their transition experience from high school to university.

(*continued*)

Table 13.2 (continued)

Topic	*Topic description*
Friendships	Friendships play an important role in the lives of young children to help them learn and develop social skills. However, you are curious to know more about the adult friendships you have made or maintained and what role they have played in your life. Investigate the roles of friendship in adulthood and how these friendships influence behavior.
Living in a multicultural society	Living in a multicultural society presents many opportunities for young adult Emirati men/women but there may also be a number of challenges to face. Investigate the perceived benefits and challenges of living in a diverse and vibrant city such as Dubai or Abu Dhabi, as experienced by Emirati men/women. Compare the perceptions of young Emiratis with those of their grandparents' generation.
Vegetarian and veganism	Vegetarianism and veganism are growing in popularity in some parts of the world. Investigate the reasons some people prefer not to eat meat or use animal products, and research the perceptions (and awareness) of Emirati young adults with regard to vegetarianism and veganism. Compare the perceptions of young Emiratis with those of their grandparents' generation.

Source: Zayed University Website, University College Course Guidelines (https://zu.libguides.com/UC/GEN185)

understanding specific research methods, finding articles, finding books, and APA citation, to name a few.

In addition to the GEN185 course and in line with the UAE's strategic plan related to education, the university's office of research launched an Undergraduate Research Scholars Program in 2010. Its main aim was to allow students to be involved in research and allow them to play a more focused and active role in the university's research portfolio. More specifically and as indicated in a ZU-URSP Executive Overview, the program seeks to:

1. Provide Emiratis at ZU with the research training, scholarly opportunities, and related experiences to be successful in research careers and graduate education
2. Promote high-quality undergraduate student research and scholarly culture

3. Increase the number of Emiratis prepared for, and entering graduate school
4. Build the capacity for Emirati citizens to conduct research on issues that are directly related to the UAE.

Research undertaken within the program can span a period of 2.5 years, or run intensively over the summer break and is faculty mentored. Faculty receive no workload adjustments and students receive no course credits toward their degree. Nonetheless, upon completion of the program, students demonstrate an ability to conduct meaningful research that is presented at local, national, and international conferences such as the American National Undergraduate Research Conference (NCUR). Travel funding is available for all papers accepted for presentation, be it in poster or presenter format and is often provided by the Office of Research. A ceremony in the format of a symposium with invited guests and speakers takes place at the end of each Undergraduate Research Scholars Program cohort. All scholars are expected to submit their manuscripts for publication to journals that are peer-reviewed for publication.

Student feedback regarding their undergraduate research experience has been nothing but positive as indicated in a summary report. For instance, one student said that, "it has opened new horizons for me. I am more interested in selecting the best graduate school after all the information I learned... I realize through this program, that research is actually important and is much more sophisticated than what I imagined it to be" (Khelifa et al. 2012).

The program is still very active today but is currently being reviewed due to several challenges. Al Hashimi and Mathews (2019) noted that students emphasized their workload to be heavy and class timings often clashed with other courses. Students also felt that they would like the course to have a more defined structure and organization. Students who were enrolled in the program but were unable to continue and graduate as an undergraduate scholar felt that it was hard to manage with their other academic tasks and they became less motivated with the heavy workload. Some students had family or transportation issues and this led to their pulling out of the program. A proposal has been submitted and is set to launch in Spring 2020. It is proposed that the program transition into an Interdisciplinary Minor in Applied Research Methods and be open to students enrolled in any major at the university. A small-scale survey showed that 84% of current students felt positively about a minor in research (Al

Hashimi and Mathews 2019). One student mentioned that "it is very important for Emirati graduates. We don't have a lot of researchers". Another student said, "It's a great idea, because we need to promote research skills in our university. Many students are lacking this skill in our university". "It was a very good idea, a minor in research methods, because every major has a research component so having a minor designated for research will prepare students for grad school and also will foster independent research and one class is not sufficient enough, you need more", is what one student had to say about having a minor in research.

One of the major benefits and added value to the newly proposed minor in applied research methods is that students will receive credit for the minor and faculty workloads will be accommodated as they teach any one of the five proposed modules that make up a total of fifteen credits (three credits per module) (see Table 13.3). A sixth and final module requires students to work independently on disseminating the research. Given the design and structure of the course, an increase in undergraduate research is anticipated once the minor is launched and given the design and structure of the courses. Moreover, the minor demands more student accountability and increased perceived value as it connects to academic and potentially accredited qualification (Al Hashimi and Mathews 2019).

Zayed University's Office of Research also provides funding for research conducted at the undergraduate level. Colleges are allocated the same amount of funding that is often used for undergraduate research-related events, student travel, and resources to support undergraduate research. For example, in Fall 2015, two College of Education faculty members, one based in Abu Dhabi and the other in Dubai, founded the College of Education Undergraduate Research and Creative Projects Symposium. The Symposium has been running every semester ever since its commencement where only a small number of students from two– to three classes participated to now having the entire College of Education

Table 13.3 Minor in Applied Research Methods Modules

Module
Principles of academic research (SP)
Research design
Data collection and management strategies (SP)
Collecting project data
RES302: Research data: Techniques for analysis and reporting

participate as part of their class work, projects, or assignments. Two of the earlier classes that were among the first to participate in the symposium were the EDC202 Human Growth and Development and EDC466 Marriage and Family classes. It was the class research projects that really helped students shape their research skills. For example, a wide range of research investigations were formed in class and driven by class discussions and student questioning.

One class project within the EDC 202 class was the "Elder's Project" that aimed at helping students to understand the perspectives of elders that they knew. Students were told that the older the person was (60 or older), the more their wisdom was to be captured in words for preservation for future generations. In order to complete this project, students were required to interview two middle/late adults. The interview questions were strategically formed in class and included both open- and close-ended questions. Upon collecting all the data, students were then asked to present their findings in a report and create a poster. A poster template was provided that included specific sections such as an abstract, introduction, participants, methodology section, data collection and analysis, results, limitations, educational implications, and discussion and conclusion. Topics that were investigated from the EDC466 Marriage and Family class included: *Father's competence regarding parenting in the UAE; Most valued qualities in a mate: Emirati female perspectives; Interracial marriage in Abu Dhabi; Showing emotions in public; Fathers' affection toward their sons in the UAE; and Arranged marriages vs. love marriages*—to name a few. Students were extremely excited to investigate these self-originated topics because they were related to their own culture and traditions and they felt that they wanted to make a positive change in their own country without jeopardizing social norms. All participating students receive certificates of participation and outstanding research or creative projects receive special awards of recognition at a closing ceremony.

Undergraduate Research at United Arab Emirates University

As the oldest university in the UAE, United Arab Emirates University has been involved in undergraduate research through mainstream curriculum research, mainstream research, and extracurricular research (McLean and Howarth 2008). McLean and Howarth (2008) described how students in the school of medicine engaged in undergraduate research and contributed to scholarship through laboratory- or community-based medical education. Students are given the option of either working with a faculty supervisor independently, or working with others as a team. An annual conference and the Gulf Co-operation Council (GCC) Medical Students' Conference provide an opportunity for students to present their work in either poster or paper format.

Furthermore, the Office of Associate Provost for Research focuses on fostering undergraduate research across the varying majors and not just the school of medicine through local and international partnerships. International partnerships include countries such as Japan, UK, USA, Australia, France, Germany, Korea, Singapore, Sudan, and Oman. Top priority is often given to research topics that are related to enhancing the UAE society (see https://www.uaeu.ac.ae). A program led by the university known as the *Summer Undergraduate Research Experiences* also serves to promote and encourage undergraduate students in research.

Undergraduate Research at New York University, Abu Dhabi

Much like United Arab Emirates University, New York University in Abu Dhabi is committed to supporting undergraduate research with a range of grants students can apply for such as the, "Grants to support students who have secured non-credit summer research positions, under the direction of a faculty supervisor; and Conference grants to support students who have been selected to present their research at academic conferences" (see https://www.uaeu.ac.ae). The grants can be used for structured undergraduate research programs, programs that have shown to really make a difference (Wilson et al. 2018).

Undergraduate Research at the American University of Ras Al Khaimah

The American University of Ras Al Khaimah offers a wide range of programs in architecture and engineering and students enrolled in any of the eight engineering programs are expected to participate in undergraduate research that is available to them across all three semesters (Fall, Spring, and Summer). An undergraduate research project course (ENGR399) is offered as a technical elective that serves many of the engineering programs. There are no set prerequisites for this course other than receiving consent from the department prior to enrolling in it. As noted on the university website, both junior and senior students engage in undergraduate research under the supervision of an engineering faculty and although the course carries three credits, a total of nine hours of research time per week is expected. Upon course completion, students either receive a pass or a fail.

Conclusion

Educational institutions across the UAE have shown interest in undergraduate research as it supports governmental efforts to provide young people with skills necessary for the future, one of which includes undergraduate research. This is demonstrated through the numerous undergraduate research engagement opportunities provided by the varying educational institutions discussed earlier in this chapter. Getting students to think outside the box and engage in critical thinking and questioning is no easy task especially in a country where reform has been at its center for the past decade. The UAE is led by leaders of great patriotism and loyalty who have stayed true to the UAE's founding leader H. H Shaikh Zayed's legacy and continue to place education as the country's top priority. Advancement in education requires research and investigations and who better to lead those investigations other than the UAE's future leaders and youth. In fact, 22-year-old Shamma bint Sohail Faris Al Mazrui was appointed Ministry of State for Youth Affairs in 2016 after a call for nominations was made. Following this, in 2017 a "Youth Hub" was formed as a community space where young professionals, entrepreneurs, programmers, media professional, and researchers could connect and share ideas. Shortly after, the Emirates Youth Council headed by the Ministry of State for Youth Affairs was formed in order to initiate youth strategies aligned with the government's current and future strategies. When it comes to strategies related to undergraduate research, an expected growth and advancement is to be seen and although funding obstacles may be evident in some countries research funding in the UAE is almost always available, particularly if research projects are aligned with the MOE strategic plans in improving the UAE or the educational system in general. From the author's personal experience of working closely for over a decade with undergraduate students, one of the biggest challenges is arguably getting students to engage in research if the research is not graded or seen as a means toward graduation. An increase in genuine interest in undergraduate research is crucial if more undergraduate research is to be demonstrated. Although the UAE has seen some outstanding undergraduate research projects, an increased number of undergraduate research programs/projects would further help the UAE's strategic plan to advance in the area of education and innovation.

Some recommendations that may further assist in the advancement of undergraduate research in the UAE may include the following:

1. Provide opportunities for students to understand the importance of engaging in undergraduate research.
2. Embed undergraduate research within a set curriculum.
3. Relate undergraduate research to the UAE culture and traditions-intrinsic motivation.
4. Provide undergraduate research opportunities across a range of classes within a major.
5. Provide both formal and informal opportunities for undergraduate students to disseminate and share their research findings.
6. Encourage students to follow up on their own research investigations while also exploring other potentially relevant topics.

References

Ahmed, A., & Abdalla Alfaki, I. M. (2013). Transforming the United Arab Emirates into a knowledge-based economy: The role of science, technology and innovation. *World Journal of Science, Technology and Sustainable Development, 10*(2), 84–102.

Al Hashimi and Mathews. (2019). Unpublished raw data.

Halaweh, M. (2019). Integrating Innovation into the UAE Qualifications Framework (QF Emirates). *Industry and Higher Education, 33*(4), 228–232.

Khelifa, M., Sonleitner, N., Wooldridge, D., & Mayers, G. (2004). Integrating research into an undergraduate family sciences program. *Journal of Family and Consumer Sciences, 96*(2), 70–71.

Khelifa, M., Feldges, B., & Jablonski, C. (2012). *ZU-URSP survey.* Unpublished raw data.

McLean, M., & Howarth, F. C. (2008). Does undergraduate student research constitute scholarship? Drawing on the experiences of one medical faculty. *Journal of the Scholarship of Teaching and Learning, 8*(1), 72–87.

MOE, UAE. (2019). *Ministry of Education Strategic Plan 2017–2021.* Retrieved from https://www.moe.gov.ae/En/AboutTheMinistry/Pages/MinistryStrategy.aspx

https://aurak.ac.ae/en/academics/school-of-engineering/undergraduate-research/

https://nyuad.nyu.edu/en/academics/undergraduate/undergraduate-research.html

Ryan, J. C., & Daly, T. M. (2019). Barriers to innovation and knowledge generation: The challenges of conducting business and social research in an emerging country context. *Journal of Innovation & Knowledge, 4*(1), 47–54.

Sarrildin, N. (2019, April 30). Abu Dhabi University awards winners of regional undergraduate research competition. *ZAWAYA*. Retrieved from https://www.zawya.com

Wilson, A. E., Pollock, J. L., Billick, I., Domingo, C., Fernandez-Figueroa, E. G., Nagy, E. S., Steury, D. T., & Summers, A. (2018). Assessing science training programs: Structured undergraduate research programs make a difference. *Bioscience, 68*(7), 529–534.

CHAPTER 14

The Status of Undergraduate Research in New Zealand: Promoted and Pervasive?

Rachel Spronken-Smith

Introduction

New Zealand's higher education system includes eight publicly funded universities, and 18 institutes of technology and polytechnics, many of which also offer undergraduate degrees. In this chapter, the focus is only on the university sector, but past research has shown that polytechnics can be very successful in promoting undergraduate research (Spronken-Smith et al. 2012; Spronken-Smith et al. 2008). All universities feature in the QS World University Rankings® 2019, with the higher education system ranked 16th in the world in the QS Higher Education System Strength Rankings 2018. New Zealand has high levels of participation in education, with 50% of New Zealanders aged 15 and over having a tertiary qualification, and 17% with a bachelor's degree or higher (Tertiary Education Commission 2014).

R. Spronken-Smith (✉)
Graduate Research School and Higher Education Development Centre, University of Otago, Dunedin, New Zealand
e-mail: rachel.spronken-smith@otago.ac.nz

N. H. Hensel, P. Blessinger (eds.), *International Perspectives on Undergraduate Research*,
https://doi.org/10.1007/978-3-030-53559-9_14

In New Zealand, the term undergraduate research tends to be used quite broadly, following the conceptualisation by Beckman and Hensel (2009) and Healey and Jenkins (2009) (Table 14.1). Thus, the term encompasses a range of curricular and extra-curricular experiences including inquiry-based learning activities as well as supervised research as part of an Honours degree or a summer research studentship. Throughout this chapter, the term undergraduate research and inquiry (URI) is used to include a range of learning experiences that develop inquiry and research skills.

New Zealand is particularly well placed to promote URI, as the Education Act 1989 included a requirement that research and teaching in New Zealand universities should be closely interdependent, and that most teaching in universities should be done by people who are active in advancing knowledge (New Zealand Government 1989). In the early 2000s, the Ministry of Education promoted "a research culture within which undergraduates learn to take a research-based approach to their lifelong educational development" (Ministry of Education 2002, p. 60). Moreover, an Academic Audit across all eight New Zealand universities in the early 2000s focused on the links between teaching and research. Although there are signals that skills produced by engagement in URI are important, the current Tertiary Education Strategy 2014–2019 (Tertiary Education Commission 2014) is not specifically calling for this pedagogical approach. For example, Priority 1—*Delivering skills for industry*, calls for the development of transferable skills including the ability to

Table 14.1 The range of dimensions of undergraduate research and inquiry-based learning

Student, process-centered	↔	*Outcome, product-centered*
Student-initiated	↔	Faculty-initiated
All students	↔	Honors students
Curriculum-based	↔	Co-curricular fellowships
Collaborative	↔	Individual
Original to the student	↔	Original to the discipline
Multi- or interdisciplinary	↔	Discipline-based
Campus/community audience	↔	Professional audience
Starting year one	↔	Capstone/final year
Pervades the curriculum	↔	Focused

Note: From Beckman and Hensel (2009, p. 40) with last two rows added by Healey and Jenkins (2009, p. 69)

communicate well, process information effectively, think logically and critically and adapt to future changes (Tertiary Education Commission 2014, p. 10). Such transferable skills are known to be well developed through URI (e.g. Murray 2017; Seymour et al. 2004; Walkington et al. 2011). The fifth priority of the Tertiary Education Strategy is *strengthening research-based institutions*. This would be a logical place to discuss the power of URI to create pathways into postgraduate research and to contribute to the research base of universities. However, this possibility has not been realised; the focus is more on staff as researchers and the need for internships and employer-linked postgraduate research.

Perhaps it is not surprising then, that URI in New Zealand has a very low profile. There are no strong central government drivers, and no national body overseeing undergraduate research (unlike the USA, the UK and Australia). In the entirely public university system, universities themselves can decide whether or not to promote and support URI. Consequently, it was necessary to undertake research to determine whether and if so, how, URI is being manifest in universities throughout New Zealand. The next section describes how data were gathered, and then key findings are presented.

Collecting Data on URI in New Zealand Universities

To determine the status of URI in New Zealand universities, the research involved a survey approach and had ethical approval from the University of Otago (D19/133). The survey began by providing a definition of undergraduate research:

> In this research, I use 'undergraduate research' as a broad term to encompass research that undergraduate students might do such as research done under an apprenticeship (e.g. undergraduate dissertations or summer studentships), research embedded in the curriculum (e.g. an inquiry-based degree program, or stand alone courses, or project work in courses), community-based research, and partnerships with research laboratories or agencies.

The questions asked about any university structures and processes to support URI including: committees; whether faculty workloads included URI; whether URI was considered in promotion of faculty; whether the university held any celebratory or showcase events for URI; and whether the university supported undergraduates to attend any conferences on URI. The survey also probed what forms of URI were implemented in and across the university, and whether there were any challenges regarding implementing URI. Finally, case studies of best practice were requested.

The survey was emailed to a network of champions for URI throughout New Zealand. Unfortunately, the network did not cover all universities, and a key problem was finding someone who actually knew about URI in their university. Surveys were sent to all eight universities, and five responded. Participants were told that the research had ethical approval, and were informed that by completing the survey, they were giving consent for their data to be used in analysis and reporting. The survey data were collated and analysed, with the findings presented below.

URI: Perhaps a Pervasive Pedagogy

All universities who responded indicated that they offer URI for students. It was apparent that URI occurs in a range of forms in all institutions. They all reported the use of the apprenticeship model, with summer studentships being very common, as well as directed studies and, for Honours students, dissertations. Summer studentships involve faculty mentoring selected undergraduates in research projects over the summer break. The funding for these projects usually comes from external research funding. The faculty mentors are typically teachers of undergraduate papers, and they view these studentships as an opportunity to get assistance on their research agenda, as well as encouraging undergraduates into postgraduate study, particularly research degrees. While these studentships tend to be offered across all disciplines, they are more common in STEM areas. In New Zealand, an Honours year can occur as part of a four-year bachelor's degree (or, in more recent years, as a separate one-year postgraduate degree). Honors degrees are offered across most disciplines. Only the higher achieving students are able to enter Honours; typically a B+ or better average for third-year papers is required. In this fourth year, the student usually takes three papers and a dissertation involving supervised research.

Respondents also reported the embedding of research in the curriculum through a variety of approaches including project work, inquiry-based learning courses or modules, and degree programmes that progressively develop inquiry and research skills. However, feedback from one university said that although many faculty would argue that URI is being progressively developed through a degree, when they tried to introduce a new degree with research at the core, it failed to be endorsed. Indeed, it is thought that there are very few degree programmes across the country that are designed in a purposeful way, using student-centered approaches to progressively develop inquiry and research skills. Most programmes would teach inquiry and research skills, but using a research-led or research-oriented approach (Griffiths 2004), where the students tend to be passive in their learning.

The exception may be Victoria University of Wellington, who, in 2011, began implementing a pan-university plan to integrate research with learning and teaching. The intent of the plan was to "ensure that students' research and inquiry skills are developed systematically in each year of their program" (Victoria University of Wellington 2011, p. 13). The plan aimed to adapt Willison and O'Regan's (2006) Research Skill Development Framework to embed research in undergraduate curricula and ensure assessment tasks include research skills. Perhaps reflecting the Victoria University policy, many examples of URI were provided, including: summer research scholarships; programmes with industry or business placements; projects at second and third year including data mining and data analysis, independent research on a chosen topic in biology, the LLB(Hons) which is largely a research degree, geography and geology field courses, inquiry-based learning in biology, chemistry, physics and engineering laboratories, and directed study courses at third year in most disciplines; community-based research such as a computer science course where students access government databases to research community issues; and partnerships with laboratories in engineering and design courses and through some joint summer scholarships.

Other forms of URI that were reported as occurring in the other New Zealand universities included community-based research, which was quite common in certain disciplines, and less commonly reported were partnerships with research laboratories or agencies (including government and non-government). The latter form of URI tended to occur through summer studentships, with students being supervised on projects that involve an external partner.

University Structures and Processes for URI

Although URI is pervasive across New Zealand universities, there is little overt support for this form of pedagogy except at Victoria University of Wellington. Indeed, it seems that perhaps close links between teaching and research are expected, and therefore it is not seen as needing support. Consequently, it was not surprising to find that none of the universities who responded to the survey had committees or offices to oversee URI, faculty workloads rarely took account of supervision involved with undergraduate research, and undergraduate research was not generally valued when faculty were applying for promotion. Three respondents commented that some faculty do discuss their promotion and use of URI in their application for teaching awards, but there is uncertainty over how this is valued given the criteria are more heavily weighted to postgraduate supervision. For example, one said I think this [the value] depends entirely on how well the proponents position it in their case (and rightly so in my view); I've certainly seen how it is valued in awards.

It was encouraging to see a plan for embedding URI in undergraduate curricula at Victoria, and the plan included specific strategies to support faculty. For example, the plan recommended a review of faculty development and promotion processes to ensure recognition of the integration of research with learning and teaching, and that faculty workloads should take account of URI. Moreover, the working party recommended that URI activities should be considered when recruiting and orienting faculty. However, although there was impetus and support for embedding URI back in 2011, over time, the championing and support for this initiative has dwindled, leaving quite patchy engagement with URI across the University.

Celebrating and Showcasing URI

None of the universities in the sample have an institution-wide celebration or symposium for URI in all its forms. One university contact reported frustration at trying to get such an event off the ground. However, three universities reported having symposia for summer students to report on their research, and all said that some disciplinary events occur. For example, one university had an Engineering Design Show that showcases undergraduate engineering research, while another had a Geography mini-conference for third-year students to report their research to an audience that includes representatives from an agency that had supplied the research problems. Two universities said they had conferences celebrating student research, in which undergraduates were able to participate alongside postgraduates.

Given the low levels of engagement with showcasing URI within institutions, it is perhaps not surprising that universities are not well engaged with supporting undergraduates to attend conferences beyond the institution. There is no national conference on undergraduate research, but New Zealand universities have been part of the *Australasian Council for Undergraduate Research* since its inception in 2012, and as such are encouraged to send students to the annual conference. So, over the last seven years there has been sporadic attendance by a few undergraduates from New Zealand universities presenting their research at the annual *Australasian Conference of Undergraduate Research (ACUR)*. The *Australasian Council for Undergraduate Research* recently lost funding by the Australian government and has become a charitable organisation, meaning that universities have to pay an annual fee to belong. Some New Zealand universities have joined and it is hoped that the hosting of *Australasian Conference of Undergraduate Research* in 2022 for the first time in New Zealand by the University of Canterbury, may encourage the remainder to join. Hosting the conference in New Zealand will certainly raise the profile of URI in this country.

Despite little engagement with the *Australasian Conference of Undergraduate Research*, there are many disciplinary national conferences that do allow undergraduates (particularly Honours students) to attend and present their research. However, not many organisations actively target undergraduate researchers, yet arguably they could do so. The main challenge is for undergraduates (or faculty) to obtain funding to support attendance by these students; funding is more readily obtainable to support attendance by graduate research students.

Challenges of Embedding URI in New Zealand Universities

One of key challenges is in defining URI and promoting a practice that is student-centred. It is apparent that many academics would say they are teaching using URI, and to a certain extent that might be true. However, the sorts of activities academics may equate with URI are essays and project work that involve students doing some research. These are not the higher-impact student-centred activities that many advocates would see as central to URI. As one respondent said, "I think some people who should know better choose not to do it, based on the old transmission practice education model." So the challenge is to support academics to shift to more activities that directly engage students in all aspects of the research cycle. A related challenge that was raised is the inconsistency in the role of URI across different degrees and even within majors of the same degree. It appears that few programmes would have URI as a core part of design with the progressive development of research and inquiry skills until senior undergraduates could undertake a piece of research. Rather, it is more common for students to learn about aspects of the research cycle—not the entire process, apart from those who do a summer studentship or an Honours degree, with a fourth year involving a dissertation.

Another key challenge, noted by two respondents, is the lack of a university body to advocate for, and support URI. One even noted a level of scepticism amongst academics that undergraduates could do research, and commented about the difficulty of getting support for initiatives from senior leaders.

Funding inequities for summer studentships was noted as a problem at one university and no doubt occurs elsewhere. The level of funding may vary from year to year, and also the distribution of the funding was variable with some disciplines getting funding in some years, but not others.

For mainstream URI courses, a particular set of challenges were identified by one respondent. These included:

- Finding enough projects for a large number of students
- Finding enough academics willing to supervise projects as it can be a high additional teaching load
- Finding projects that are doable in a semester by students who lack research experience
- Trying to ensure projects are of a similar scope and size
- Ensuring the main focus is on student learning, not staff publications, and that students are not viewed as research technicians.

In addition to these challenges is the need for ethical approval for some projects and the time this can take. A final challenge is the need to support academics to teach using an inquiry-based learning approach (e.g. Spronken-Smith and Harland 2009). While educational development units take on this role, it was only one university—Victoria—that had an institutional mandate to explicitly promote and support such approaches.

URI in Practice

In 2008–2009, a nationally funded project explored 14 examples of URI at tertiary institutions throughout New Zealand (Spronken-Smith and Walker 2010; Spronken-Smith et al. 2011a; Spronken-Smith et al. 2008). The project uncovered some excellent practice—mainly at the level of a course or module. Examples of inquiry at universities were documented in Communications Disorders, Engineering and Sociology (University of Canterbury), History, Psychology, Business and Architecture (Victoria University of Wellington) and Ecology, Endocrinology and Political Communication (University of Otago). The project also uncovered an excellent example of redesigning a degree programme to take an inquiry approach. The Bachelor of Science degree in Ecology at Otago was purposefully overhauled to progressively build inquiry and research skills from first to third (final) year students (Spronken-Smith et al. 2011b). Table 14.2 shows how URI activities were used at each year level. In the first year ECOL111 paper, students undertake a small-scale research project in a bush area near the University, developing their ability to take observations, develop and test a hypothesis, and communicate their findings. In the ECOL211 paper, students work in groups, guided by a tutor, to undertake research on an ecosystem, developing posters about the

Table 14.2 Embedding of undergraduate research and inquiry activities into core papers of the ecology degree at the University of Otago

Course	*Nature of inquiry activity*
ECOL111 Ecology and Conservation of Diversity	Field and laboratory project on invertebrate biodiversity with a research report
ECOL211 Ecology of Communities and Ecosystems	Research project on an ecosystem with creation of a poster and a research proposal
ECOL212 Ecological Applications	Reviewing and discussing approaches to key ecological questions
ECOL313 Ecology Field Course	Generation and execution of research project, with findings communicated in the style of an academic journal article

Note: Adapted from Spronken-Smith, Walker, Dickinson, Closs, Lord et al. (2011b, p. 727)

ecosystem and then developing a research proposal based on a disturbance to the ecosystem. In ECOL212, classes were taught by faculty critiquing their own research and students had to critique a major theoretical ecological question. Finally, in the capstone ECOL313 paper, students had to design a research project that was implemented during an eight-day residential fieldtrip. Subsequently, students analysed their data and wrote up their findings in a style suitable for submission to an academic journal (and several have gone on to publish).

A spin-off from this nationally funded project was some professional development sessions to support a transition to URI. Subsequently, faculty development work occurred with curriculum teams in various institutions around New Zealand, and this has resulted in more programmes adopting URI approaches. In the collection of data from the universities for this chapter, one of the case studies offered was a course that developed following one of these professional development sessions: Integrative Studies in Veterinary Sciences at Massey University.

The Integrative Studies course occurs in the third year of the degree (the programme is 5 years) and has a duration of one semester (12–13 weeks). Approximately 125 students are involved and in the first week they are introduced to two frameworks that can be used to develop a research proposal, as well as six possible research topic areas. In week two, in an activity known as "The Hunger Games", students had to sort themselves into groups of four to six, according to the topics of most interest. Students then had a workshop on using the library and critical evaluation of the literature and in week three had to complete a critical evaluation of a paper related to their research topic. For the next six weeks they worked with their mentors and had workshops on study design, ethics and scientific writing. Online resources were also available to support their learning about research methods. In week 10, they presented their proposal and gained feedback on a written draft proposal from the course coordinator and their mentor. Over the next two weeks, they revised their proposal and submitted the final version for assessment. Some of the proposals are then used for summer research projects, ultimately resulting in some publications.

Another example mentioned in the survey is a third-year capstone course in Geography at the University of Canterbury. In 2001, this course was redesigned to take an inquiry approach, with a local agency providing real-world problems for student groups to tackle. The course attracts 50–70 students, and they opt into groups of four to six to study problems of interest. Lectures are used sparsely, with most learning occurring through a weekly group meeting with mentors, and workshops on research methods and skills. Students need to identify a research question related to the problem and then design and implement a study to tackle this question. They present their findings via a written report, and oral and poster presentations at a mini-conference, which members of the external agency attend near the conclusion of the course. This course has been researched with Spronken-Smith (2005) detailing the nature of the course and student and faculty experiences, and Spronken-Smith and Harland (2009) discussing how to support faculty to transition to this teaching approach.

A further example, quite novel, is the *Matariki Undergraduate Research Network* (MURN) (Spronken-Smith et al. 2018). The *Matariki Network* comprises seven universities across the globe, and this MURN initiative involved the University of Otago, NZ, alongside the University of Western Australia, Durham University in the UK and Queen's University, Canada. For both 2012 and 2013, three of these universities recruited undergraduates to undertake research on internationalisation. The students were taught about higher education research in a global classroom using web conferencing to link the students across countries, and then they were mentored on local research projects. Over the two years, although the cohorts reaped the usual benefits of being involved in undergraduate research, the global connections were not as strong as had been hoped. Really, students needed to be working on collaborative cross-country projects, although this would have meant some difficult navigation of ethics approval systems. Unfortunately, the initiative was not sustainable as it relied on substantial funding. However, the proponents felt that if it was made a credit bearing course, then it could be sustainable. It was certainly a powerful learning experience for the students, and when some (University of Otago and University of Western Australia) met at the *Australasian Conference of Undergraduate Research* in 2013, they immediately bonded with classmates they had previously only met online.

As mentioned earlier, one of the main forms of URI at New Zealand universities is summer studentships (or summer research scholarships). As an example, at the University of Waikato, these are available for students enrolled in an undergraduate degree, a final-year Honours degree or the first year of a taught master's degree. The studentship programme typically involves about 80 students over a period of 10 weeks. The studentships are intended to support and enhance research at the University, and, importantly, to assist students—particularly those considering further study—in the development of their research skills and to experience the challenges and rewards of research work. Many studentships result in research publications, co-authored with mentors.

Conclusions

As indicated by the subtitle of this chapter, there is a question about whether URI is promoted and pervasive. Certainly, the Education Act is a strong promoter of close links between teaching and research, and at the turn of the century, the Ministry of Education was trying to advocate for research-based learning, with undergraduates developing research skills. More recent government policies, despite an emphasis on graduate outcomes, have not been pushing so explicitly for research skill development.

The universities in New Zealand all have some imperatives regarding research-informed or research-led teaching, but few go further to explicitly advocate for URI. By virtue of offering Honours degrees, and because of the widespread nature of summer research scholarships, many undergraduates will be exposed to URI. However, to foster research and inquiry skills in all undergraduates requires an institutional commitment, with policy promoting and supporting this, and structures in place to enable the embedding of URI in curricula. The Victoria University of Wellington is the only institution in the sample to have a plan to integrate research with learning and teaching, but although this was initially well supported, over time the initiative has lost impetus. The Victoria example shows that with an institutional imperative and educational development support, URI can be embedded across an institution, but this support needs to be ongoing. Fortunately, many examples of URI have been embedded across programmes, but the implementation is not even across the institution. For the other universities included in this study, practices were very patchy. Some excellent URI initiatives are occurring, but nowhere is there systematic embedding of URI across the institution.

Undergraduate research and inquiry is not celebrated at an institutional level, with the exception of summer research projects, and nor is there a national conference for URI. Fortunately, there is the *Australasian Conference on Undergraduate Research*, but logistical issues (costs and timing) often preclude students from New Zealand participating. Worse, this involvement might have stifled the development of a national conference that could perhaps have elevated the status of URI in New Zealand. But all is not lost. There is evidence that URI is quite pervasive, providing a terrific platform for further development. The *Matariki Undergraduate Research Network* showed that URI can be conducted in a global classroom, linking undergraduates across the world. To sustain such initiatives, they need to be credit-bearing programmes, and there is great potential for collaborative cross-border undergraduate research. An exciting new initiative under development is "Posters in the Beehive" (the building for New Zealand's national government), similar to events such as "Posters on the Hill" (USA) or "Posters in Parliament" (UK), as a way to showcase URI in New Zealand, and it is thought there could be widespread support from across the university sector for such an event.

References

Beckman, M., & Hensel, N. (2009). Making explicit the implicit: Defining undergraduate research. *CUR Quarterly., 29*(4), 40–44.

Griffiths, R. (2004). Knowledge production and the research-teaching nexus: The case of the built environment disciplines. *Studies in Higher Education, 29*(6), 709–726.

Healey M., & Jenkins, A. (2009). *Developing undergraduate research and inquiry.* Research Report to the Higher Education Academy, York: Higher Education Academy. Available from https://www.heacademy.ac.uk/knowledge-hub/developing-undergraduate-research-and-inquiry. Accessed 1 July 2019.

Ministry of Education. (2002). *Tertiary education strategy 2002/07.* Wellington: Ministry of Education.

Murray, J. L. (2017). *Undergraduate research for student engagement and learning.* New York: Routledge.

New Zealand Government. (1989). *The Education Act.* Available at: http://www.legislation.govt.nz/act/public/1989/0080/latest/DLM175959.html. Accessed 1 July 2019.

Seymour, E., Hunter, A.-B., Laursen, S. L., & DeAntoni, T. (2004). Establishing the benefits of research experiences for undergraduates in the sciences: First findings from a three-year study. *Science Education, 88*(4), 493–534.

Spronken-Smith, R. A. (2005). Implementing a problem-based learning approach for teaching research methods in geography. *Journal of Geography in Higher Education, 29*(2), 203–221.

Spronken-Smith, R. A., & Harland, A. (2009). Learning to teach with problem-based learning. *Active Learning in Higher Education, 10*, 138–153.

Spronken-Smith, R., & Walker, R. (2010). Can inquiry-based learning strengthen the links between teaching and disciplinary research? *Studies in Higher Education, 35*(6), 723–740.

Spronken-Smith, R.A., Walker, R., Batchelor, J., O'Steen, B., Angelo, T., & Matthews, H. (2008). *Inquiry-based learning*. Prepared for the New Zealand Ministry of Education, July 2008.

Spronken-Smith, R. A., Walker, R., Batchelor, J., O'Steen, B., & Angelo, T. (2011a). Enablers and constraints to the use of inquiry-based learning in undergraduate education. *Teaching in Higher Education, 16*(1), 15–28.

Spronken-Smith, R. A., Walker, R., Dickinson, K. J. M., Closs, G. P., Lord, J. M., & Harland, T. (2011b). Redesigning a curriculum for inquiry: An ecology case study. *Instructional Science, 39*(5), 721–735.

Spronken-Smith, R. A., Walker, R., Batchelor, J., O'Steen, B., & Angelo, T. (2012). Evaluating student perceptions of learning processes and intended learning outcomes under inquiry approaches. *Assessment and Evaluation in Higher Education, 37*(1–2), 57–72.

Spronken-Smith, R. A., Sandover, S., Partridge, L., Fawcett, T., & Burd, L. (2018). The challenges of going global with undergraduate research: The Matariki Undergraduate Research Network. *SPUR: Scholarship and Practice of Undergraduate Research, 2*(2), 64–72.

Tertiary Education Commission. (2014). *Tertiary education strategy 2014–2019*. Wellington: Ministry of Education and the Ministry of Business, Innovation and Employment.

Victoria University of Wellington. (2011). *Review of undergraduate research*. Unpublished report.

Walkington, H., Griffin, A. L., Keys-Mathews, L., Metoyer, S. K., Miller, W. E., Baker, R., & France, D. (2011). Embedding research-based learning early in the undergraduate geography curriculum. *Journal of Geography in Higher Education, 35*(3), 315–330.

Willison, J., & O'Regan, K. (2006/2013). *Research skill development framework*. Available at: https://www.adelaide.edu.au/rsd/framework/rsd7/. Accessed 1 July 2019.

CHAPTER 15

Transforming Undergraduate Research at Canadian Universities

Brad Wuetherick

Introduction

> The human spirit thrives on discovery. We must integrate discovery into all aspects of learning. The "Great University" of the twenty-first century must involve students in exploring our grand challenges. (Samarasekera 2005, p. 4).

Undergraduate research has been an important part of the Canadian higher education landscape for several decades, and a significant proportion of students would have some opportunities to experience research over the course of their undergraduate education, yet the current state of undergraduate research in Canadian higher education is highly varied. The leadership of every single university in Canada would likely agree that undergraduate research is one of the high-impact practices that they must offer as part of their learning environment (Kuh 2008). And each Canadian university would have examples of mentored and curriculum embedded

B. Wuetherick (✉)
Learning and Teaching, Dalhousie University, Halifax, NS, Canada
e-mail: brad.wuetherick@dal.ca

N. H. Hensel, P. Blessinger (eds.), *International Perspectives on Undergraduate Research*,
https://doi.org/10.1007/978-3-030-53559-9_15

experiences where some proportion of their undergraduate students experienced undergraduate research over the course of their degree programs. Very little is known systematically, however, about the current state of undergraduate research in Canada.

There are no consistent institutional approaches to coordinating undergraduate research activities (e.g., through the creation of offices for undergraduate research as has occurred widely in the US), if any institutional approach is even undertaken. There is no national organization focused on advocating for undergraduate research or coordinating national conversations about advancing undergraduate research. There have been a few attempts to create such an organization, but they have not been able to be sustained. As was articulated by Vajoczki (2010, p. 41), "much of what is known about undergraduate research (in Canada) is anecdotal, unstandardized, and uneven. Published or other readily accessible information and evidence is relatively scarce." This assessment would largely hold true today.

Within that context, however, it is important to emphasize that undergraduate research experiences have become a ubiquitous part of the higher education landscape across the country, significantly mirroring the conversations happening in many other jurisdictions (Larson et al. 2018). And research and inquiry skills are highly sought by employers across the country, though in many recent studies they are implicit rather than explicit within the list of attributes articulated by employers (Conference Board of Canada 2019; Lennon 2010).

After a few decades where informal and formal mentored research opportunities were available for relatively small groups of high-achieving students, there was a move toward the inclusion of research and inquiry in the undergraduate curriculum in the 1990s and early 2000s. This was particularly championed at McMaster University, where open-ended, inquiry-based learning was first implemented starting in 1981 in their innovative Arts and Science Interdisciplinary Inquiry program, and subsequently across several of their undergraduate faculties (Jenkins et al. 2004; Justice et al. 2007a, b; Vajoczki 2010; Marquis 2017).

The Canadian conversation about undergraduate research did not gain significant, widespread traction, at least at the level of institutional academic planning, until the turn of the millennium, particularly following the high-profile Boyer Commission (1998) in the US articulated a vision for higher education where research-based learning was expected across the disciplines. While Canadian institutions did not participate in the

Boyer Commission process, there was widespread realization that virtually all Canadian universities would struggle to demonstrate that their institution met the expectations laid out by the commission's final report. The conversation began to accelerate in the mid-2000s, particularly surrounding two national summits on the integration of research, teaching, and learning, hosted in 2005 and 2006 at the University of Alberta (Wuetherick 2007; Wuetherick and McLaughlin 2011). It was in this period that the then-incoming President of the University of Alberta, Indira Samarasekera, as represented by the quote that starts this chapter from her inaugural Presidential address, incorporated students' involvement in "discovery" as a core element of her vision for the institution, mirroring similar strategic commitments at several other universities across the country (Samarasekera 2005). This is also when scholarship about undergraduate research in Canadian universities began to appear within the broader scholarly discourse about undergraduate research (e.g., see Bartlett 2003 or Hoddinott and Wuetherick 2006).

Defining Undergraduate Research in Canada

Unlike the definition provided by the Council for Undergraduate Research (2015) in the US, Canadian higher education has not had a unifying definition that informs institutional approaches to undergraduate research. Generally, at Canadian universities undergraduate research is understood to exist on a number of continua in line with those articulated by Beckman and Hensel (2009). Many institutions in Canada would use holistic language around "undergraduate research and inquiry," or "undergraduate research, scholarly and creative activities," where students are, to use the language of Brew and Boud (1995), engaged in research into the "commonly known" (commonly known by faculty in a discipline, but new to the student), the "commonly unknown" (known only by a few specialists in a field, but unknown to the student as well as faculty outside of that specialty), and the "totally unknown" (and therefore makes an original contribution to the field).

Research undertaken at the University of Alberta, for example, explored how students conceptualized the integration of research into the learning environment, and the student respondents gave examples of learning about research (particularly the research conducted by the faculty on campus), learning about research methods (in methodology classes), learning through research and inquiry experiences (embedded within courses), and undertaking mentored research with a faculty mentor (Wuetherick 2007; Turner et al. 2008). The proportion of students experiencing each of these modalities decreased along the continuum, though the majority of students agreed or strongly agreed that they learn best through research or inquiry experiences (Turner et al. 2008; Wuetherick and McLaughlin 2011).

For the Canadian higher education context, it is true that "there is no one correct definition … institution(s) will best access the many benefits of undergraduate research by carefully formulating a definition or definitions that fit its campus culture and its unique institutional mission" (Beckman and Hensel 2009, p. 44). A lack of a unifying national definition, however, has not impacted the widespread implementation of undergraduate research experiences across most campuses.

Mentored Undergraduate Research

Mentored undergraduate research, defined as a research experience where an undergraduate student works intensively with a research mentor (most often a faculty member), takes many forms in Canadian universities. Most institutions in Canada have well-established honors degree programs where students would undertake some form of mentored research as a formal part of their undergraduate degree programs, normally resulting in an honors thesis or dissertation. The proportion of students who complete honors programs is below 20% nationally, and within many disciplines would be lower than 10%. Other academic programs have required research experiences or capstone courses that require students (whether individually or in groups) to undertake a mentored research project as a requirement for the completion of their undergraduate program. For example, through Dalhousie University's Research in Medicine program, every student in the medicine program completes a mentored research project as an embedded requirement of the degree, including (at a minimum) disseminating the results of the research through a presentation at a research conference (Wuetherick et al. 2018). Mentored research with faculty, however, often occurs outside of the formal curriculum at Canadian universities.

Undergraduate students have been undertaking mentored research informally for many decades, most often through funded summer research assistantships where students work with researchers for up to 16 or more weeks over the summer semester. Much like the history of the National

Science Foundation in the US, these mentored research experiences were formalized when the Undergraduate Student Research Awards (USRAs) were established in 1980 by the National Science and Engineering Research Council. Ever since, Canada has had a well-established tradition of mentored, summer, undergraduate research opportunities at universities across the country. The number and types of these funded, mentored research experiences, however, have been highly skewed to the STEM disciplines. Mentored experiences in the non-STEM fields have not had sustained and targeted funding, though many faculty across the disciplines would use research grant funding to support undergraduate students as researchers on their projects and many institutions invest their own resources to provide opportunities outside of the STEM disciplines. Most mentored research experiences are provided by full-time continuing faculty, though that depends highly on the individual contexts across and within institutions.

These mentored research experiences are normally one-on-one mentored experiences between undergraduate students and researchers, though the nature of the mentorship offered varies greatly depending on the disciplinary and institutional contexts. At the larger research universities, for example, students are much more likely to work with undergraduate peers while undertaking their projects and to be supported by graduate student and postdoctoral fellow mentors in addition to their faculty mentor (particularly in the STEM disciplines where larger, collaborative research teams are common).

There have been almost no formal attempts to estimate the proportion of students who experience some form of mentored research. That said, the National Survey of Student Engagement, which the vast majority of Canadian universities administers at least once every three years, reports that for the years 2017–2018 about 22% of final year undergraduate students in Canada agreed that they had at least one experience conducting research with faculty, which is slightly down from 23% of final year students in 2014 (NSSE 2019). The proportion of students reporting a research experience through NSSE, however, varies significantly by institution type—where at the group of 15 medical-doctoral research universities (the U15) the proportion climbs to 26% of final year undergraduate students, as well as by discipline—from as low as 9–13% in business, education, communication/public relations and other social service professions, to as high as 33% in the physical and mathematical sciences and 42% in the biological and agricultural sciences. This estimate is likely on the high side, however, as there is no definition provided on the NSSE survey about what is intended to be included as "research with faculty," so students may be reporting course-based undergraduate research experiences that would not normally be considered mentored undergraduate research. NSSE also reports that students historically underserved by higher education, including mature, first-generation, part-time, and transfer students, all reported a lower likelihood of experiencing research with a faculty member during their undergraduate learning environment even though the literature has confirmed that these experiences are more beneficial for those students (NSSE 2019; Kuh 2008).

One unique Canadian initiative to support mentored undergraduate research is coordinated by Mitacs, a national, not-for-profit organization (Mitacs 2019). Mitacs coordinates a Globalink research awards program to enable close to 400 undergraduate students studying at Canadian universities to travel overseas to undertake a 12- to 24-week research internship at a partner institution. The Globalink program also funds close to 800 undergraduate students studying outside of Canada to travel to one of 45 university partners within Canada to undertake a 12-week research internship. The Mitacs program also funds significant research training at the graduate and postdoctoral levels. Over its close to 20-year history, it has funded a total of 20,000 research internships at the undergraduate, graduate, and postdoctoral levels (Mitacs 2019).

There are also several interesting initiatives at Canadian universities explicitly targeting historically underserved students to have access to mentored undergraduate research experiences. For example, Imhotep's Legacy Academy (2019) at Dalhousie University, which is an initiative designed to increase the proportion of African Nova Scotian students entering STEM fields, has targeted summer research fellowships that allow African Nova Scotian students to participate in a mentored summer research experience. Another award-winning initiative is called the *Knowledge Makers* program at Thompson Rivers University, where 15 Indigenous students from disciplines across the institution are partnered with researchers, Indigenous elders, and community members to "make knowledge" through a multimodal approach that weaves Indigenous and Western knowledges to support Indigenous research and researcher development (Naepi 2019; Naepi and Airini 2019).

Scaling Undergraduate Research in the Curriculum

Canadian universities have, for several decades, undertaken various models of undergraduate research embedded within the curriculum. This chapter has already discussed embedding research experiences in honors programs or through capstone research experiences, which would be the most common manifestation of curriculum-embedded undergraduate research in Canada. Most Canadian universities, however, are trying to tackle the challenge laid out in the Boyer Commission report (1998), and reiterated by Healey and Jenkins (2009), to make research-based learning the norm so that all undergraduate students in all disciplines have an authentic research experience. This challenge requires all academic programs to explore how authentic course-based undergraduate research experiences can be scaled to large numbers of undergraduate students. Though there has not been systematic evaluation of the proportion of students who undertake a curriculum-embedded research experience, it is certainly higher than the proportion of students who responded on NSSE that they had experience in "research with faculty." There is significant evidence from research on undergraduate research and inquiry at Canadian institutions that learning through research has improved academic outcomes with respect to disciplinary knowledge and skills as well as the development of human skills, such as communication, information literacy, ethical judgment, and critical reasoning (Justice et al. 2007a, b; Turner et al. 2008). While there are numerous Canadian examples of course-based or curriculum-embedded undergraduate research, this chapter highlights examples from three different institutions with whom the author has direct experience as an employee and collaborator that demonstrate a range of ways in which Canadian universities are scaling undergraduate research in the curriculum.

Many universities have created explicit research project courses at their institution that provide students with a one or two semester mentored research experience in the context of a credit course. For example, several Science departments at the University of Alberta (UofA) have created a suite of one or two semester courses (such as: Bio 298, 299, 398, 399, 498, and 499) that students can take to gain hands-on authentic research experiences over the course of the degree, with the expectations for the level of student autonomy and quality of research work increasing from second year to fourth year of study. Some of those same departments at the UofA have also created a stand-alone Research Certificate in Science (in Biology, Psychology, and the Biomedical Sciences) that enables students who have completed seven semesters of explicit coursework related to research, including courses focused on data handling and practical research skills/techniques in addition to the research project courses, plus have presented their research at one or more conferences on or off campus, to receive a research certificate in addition to their Bachelor's degree in their major field of study (University of Alberta 2019).

Another common model for scaling undergraduate research is to redesign existing courses to change the pedagogical and assessment strategies to enable an authentic research and inquiry experience. For example, at Dalhousie University, in the Department of Chemistry, two faculty members partnered with a senior undergraduate student to completely transform a second-year physical chemistry lab, replacing the previous "cookie-cutter" labs with inquiry-based experiences that reinforced physical-chemistry concepts such as thermodynamics and kinetics and introduced research skills that were applicable to academic or industrial energy-storage research (Licht et al. 2018). Another example, though this course has subsequently changed its focus again, was the Animal Science 200 class at the University of Alberta's Department of Agricultural, Food and Nutritional Sciences. This course, which was the introduction to animal agriculture required for all students in the BSc in Agriculture, was revamped to group students into collaborative teams where the students had to research science-based answers to quirky questions about animal agriculture. The initiative, which was called *There's a Heifer in Your Tank* after a particular signature question one group of students had to research related to methane capture and methane-powered vehicles, ended up attracting significant interest and external funding from outside the institution, with the students' projects being featured in the local newspaper, on a website, and in other dissemination formats. The final project presentations, which often involved significant creativity on the part of the students, were presented in front of hundreds of friends, family members, and community/industry leaders (Robinson et al. 2006).

Increasingly, many faculty are redesigning courses to partner with community organizations with students to undertake community-based undergraduate research. In the Faculty of Computer Science at Dalhousie University, students take a course where each year a different community partner identifies an information-technology-related issue that they are trying to grapple with, and that they do not have the internal capacity to resolve. The students work in groups to research, design, and implement the solution to the problem (Blouin 2019). This model for undergraduate research can move beyond the curriculum as well. At the University of Saskatchewan, for example, a faculty member in the College of Arts and Science has created a community-engaged collaboratorium where community partners identify research projects for which students in the humanities and social sciences are then hired to conduct research. Projects have included collecting and analyzing Indigenous histories as a means of contributing to justice and social change within museums or developing a database of colonial history in the province that will be used to inform the overrepresentative population of Indigenous people in the legal system (USask 2019a).

Arguably, the most comprehensive and innovative approach to scaling undergraduate research in Canada is the *First Year Research Experience* (FYRE) project at the University of Saskatchewan (2019b). A partnership of the Undergraduate Research Initiative of the Office of the Vice-President Research and the Gwenna Moss Centre for Teaching and Learning, FYRE pairs an educational developer with course design expertise together with individual faculty members to redesign first-year courses to provide students with authentic research experiences. Through FYRE, first-year students are expected to develop research skills through what they call the research arc—develop a question; investigate that question using appropriate disciplinary methodologies; and share the findings with an audience beyond the professor. Since 2014, there have been 8262 undergraduate students enrolled in 27 different FYRE classes, including 1063 students in Agriculture 111 (Discovery in Plant and Soil Sciences); 800 students in Geography 120 (Introduction to Global Environmental Systems); 767 students in Sociology 111 (Foundations in Sociology: Society, Structure, and Process); and 402 students in Commerce 104 (Business Statistics) (USask 2018). The initiative is part of the broader Undergraduate Research Initiative that is supporting the transformation of the undergraduate student experience at the University of Saskatchewan. In their recent annual report, the URI reported that out of 3035 students enrolled at the university starting in 2014, more than 50% had engaged in an authentic, course-based research experience by 2018, including almost 25% of those students experiencing research in their first year of study (USask 2018).

Supporting the Undergraduate Research Environment

Canadian universities have invested significantly in supporting the undergraduate research environment to ensure students and faculty have the most beneficial outcomes for both mentored and curriculum-embedded undergraduate research. Many universities have created undergraduate research offices or initiatives, but there are a number of different models in use across the country. Undergraduate research offices have been set up as stand-alone centers (e.g., the University of Ottawa Centre for Research Opportunities—https://research.uottawa.ca/centre-research-opportunities/). Others have been affiliated with the Student Affairs portfolio (e.g., the University of Alberta Undergraduate Research Initiative—https://www.ualberta.ca/undergraduate-research-initiative). A third model is for the office to be part of the Office of the VP Research (e.g., the University of Saskatchewan's Undergraduate Research Initiative—https://vpresearch.usask.ca/students/undergraduate/undergraduate-research.php). A fourth model is for the undergraduate research office to be part of the teaching and learning center (e.g., the College of Creativity, Discovery and Innovation at the Taylor Institute for Teaching and Learning at the University of Calgary—https://taylorinstitute.ucalgary.ca/students/undergraduate). And finally, other institutions have relied on faculty-level coordination for undergraduate research (e.g., the Faculty of Science at McGill University—https://www.mcgill.ca/science/research/undergraduate-research). For those institutions that have created some form of central coordination for undergraduate research, these units are often expected to support the distribution of funding to undergraduate researchers, coordinate central events or journals for undergraduate researchers, as well as champion the implementation of undergraduate research in the curriculum.

Most Canadian universities are also actively supporting the dissemination of undergraduate research through conferences and peer-reviewed journals. Ensuring that undergraduate students have the opportunity to disseminate their work in authentic environments has been identified as a key component of an authentic research experience (Spronken-Smith et al. 2013). Undergraduate research journals have been created at the level of disciplines (e.g., *The Dalhousie Medical Journal*—https://ojs.library.dal.ca/DMJ), at the level of the institution (e.g., the *University of Saskatchewan Undergraduate Research Journal*—https://usurj.journals.

usask.ca/), and at the national level (e.g., the *Canadian Journal for Undergraduate Research* at UBC—https://www.uroubc.com/cjur/). Similarly, undergraduate research conferences have been established at the level of departments or faculties (e.g., the Faculty of Science Undergraduate Research Conference at McGill University—https://www.mcgill.ca/science/research/undergraduate-research/urc), the institutional level (e.g., the Festival of Undergraduate Research and Creative Activities at the University of Alberta—https://www.ualberta.ca/undergraduate-research-initiative/furca), and the national level within the disciplines (e.g., the Canadian Undergraduate Computer Science Conference—http://www.cucsc.ca/, or Undergraduate Neuroscience Conference—https://canada-unc.com/). Unlike the US (NCUR), Australia (ACUR), or the UK (BCUR), Canada does not have a national cross-disciplinary undergraduate research conference.

Perhaps the most important, but often overlooked, support for undergraduate research is in the form of support for undergraduate research mentorship. The recognition of undergraduate research mentorship as a component of faculty workload, and subsequently in processes such as tenure and promotion, varies significantly both within and between institutions. Recent research has highlighted the critical importance that mentors play on the development of not only the undergraduate students' academic and research identity, but also on the intersection of that identity with their personal, sociocultural identities (Palmer et al. 2015, 2018a, b). Efforts to support the development of effective mentors not only impacts one-on-one mentored research experiences, but also has a profound impact on the course-based and curriculum-embedded experiences as well (Wuetherick et al. 2018), yet faculty mentors have often received little to no training or support in how to be an effective mentor inside or outside of the classroom. There have been calls for more systemic support for evidence-informed mentorship skill development to support student success through research experiences at the undergraduate as well as the graduate levels (Lunsford et al. 2017).

Students Driving Change

At many Canadian universities, the undergraduate students have championed aspects of undergraduate research at their respective institutions. Many of the undergraduate research conferences or journals that exist in universities across the country are driven by undergraduate student volunteers. At the University of Alberta, the Undergraduate Research Initiative (the office formally tasked with coordinating undergraduate research on campus) owes its creation to advocacy by the Students' Union, who made its establishment one of their highest priorities (Wuetherick and McLaughlin 2011). The UofA Students' Union also helped drive several research projects exploring undergraduate students' awareness of, and experiences with, research as part of an institutional commitment to transform the undergraduate student learning environment (Wuetherick and McLaughlin 2011). Students driving change in the area of undergraduate research is perhaps best represented by the University of British Columbia's student group called Undergraduate Research Opportunities (URO 2019). The URO student group coordinates the *Canadian Journal of Undergraduate Research* (mentioned above), organizes events and conferences throughout the academic year related to undergraduate research, coordinates a research mentorship program for new student researchers, provides resources about undertaking and disseminating research, and disburses research and travel awards to students based on resources they have attracted from sponsors on and off campus.

Another area of recent focus within Canadian universities, championed in particular by the MacPherson Institute at McMaster University, is the involvement of students-as-partners in teaching and research. While students-as-partners initiatives manifest in a number of curricular and course design processes that would not fall under an umbrella of undergraduate research, there are many examples of students-as-partners projects with faculty, not just within subject-specific disciplinary research but also in the area of the scholarship of teaching and learning (Healey et al. 2014; Cook-Sather et al. 2014; Marquis 2017). While the students-as-partners movement is global, within Canadian universities it reflects a slowly shifting orientation to a recognition of students-as-partners in the scholarly, knowledge-building community across the entire spectrum of the teaching–research nexus within universities.

Conclusion

As demonstrated throughout this chapter, there is a significant amount of undergraduate research undertaken in Canadian universities, though the largest challenge remains scaling undergraduate research opportunities to enable all students to have an authentic experience with research during their program of studies. The Canadian landscape has continued to evolve over the past few decades as institutions increasingly strive to ensure that learning through research becomes increasingly the norm across all disciplines. There remains, however, significant room for all higher education institutions in Canada to enhance their undergraduate research activities, particularly if institutions espouse, as I suspect many do, that all students should have an authentic research experience. Indeed, it is in part what makes higher education *higher* (Healey and Jenkins 2009).

References

Bartlett, K. (2003). Towards a true community of scholars: Undergraduate research in the modern university. *Journal of Molecular Structure (THEOCHEM), 666–667*, 707–711.

Beckman, M., & Hensel, N. (2009). Making explicit the implicit: Defining undergraduate research. *CUR Quarterly, 29*(4), 40–44.

Blouin, C. (2019). Personal communication.

Boyer Commission on Educating Undergraduates in the Research University. (1998). *Reinventing undergraduate education: A blueprint for America's research universities.* Stony Brook: State University of New York at Stony Brook.

Brew, A., & Boud, D. (1995). Teaching and research: Establishing the vital link with learning. *Higher Education, 29*, 261–273.

Conference Board of Canada. (2019). *Employability skills toolkit.* Toronto: Conference Board of Canada. Retrieved from: https://www.conferenceboard.ca/edu/toolkit.aspx

Cook-Sather, A., Bovill, C., & Felten, P. (2014). *Engaging students as partners in learning and teaching: A guide for faculty.* San Francisco: Jossey-Bass.

Council for Undergraduate Research. (2015). *About CUR.* Council for Undergraduate Research. Retrieved from http://www.cur.org/about_cur

Healey, M., & Jenkins, A. (2009). *Undergraduate research and inquiry.* York: Higher Education Academy.

Healey, M., Flint, A., & Harrington, K. (2014). *Engagement through partnership: Students as partners in learning and teaching in higher education.* London: Higher Education Academy.

Hoddinott, J., & Wuetherick, B. (2006). The teaching-research Nexus. *Education Canada, 46*(1), 32–35.

Imhotep's Legacy Academy. (2019). *About ILA.* Retrieved from: https://www.dal.ca/faculty/science/imhotep/about.html

Jenkins, H., Ferrier, B., & Ross, M. (Eds.). (2004). *Combining two cultures – McMaster University's arts and science programme: A case study.* Lanham: University Press of America.

Justice, C., Rice, J., Warry, W., Inglis, S., Miller, S., & Sammon, S. (2007a). Inquiry in higher education: Reflections and directions on course design and teaching methods. *Innovative Higher Education, 31*(4), 201–214.

Justice, C., Rice, J., Warry, W., & Laurie, I. (2007b). Taking an 'inquiry' course makes a difference: A comparative analysis of student learning. *Journal on Excellence in College Teaching, 18*(1), 57–77.

Knowledge Makers. (2019). *About Knowledge Makers.* Retrieved from: https://www.tru.ca/indigenous/knowledge-makers-tru/about.html

Kuh, G. (2008). *High-impact educational practices: What they are, who has access to them, and why they matter.* Washington, DC: Association of American Colleges and Universities.

Larson, S., Partridge, L., Walkington, H., Wuetherick, B., & Moore, J. (2018). An international conversation about mentored undergraduate research and inquiry and academic development. *International Journal for Academic Development, 23*(1), 6–14.

Lennon, M. C. (2010). *A fine balance: Supporting skills and competency development.* Toronto: Higher Education Quality Council of Ontario.

Licht, F., Aleman Milan, G., & Andreas, H. (2018). Bringing real-world energy-storage research into a second-year physical chemistry lab using a MnO_2-based supercapacitor. *Journal of Chemistry Education, 95*(11), 2028–2033.

Lunsford, L., Crisp, G., Dolan, E., & Wuetherick, B. (2017). Mentoring in higher education. In D. Clutterbuck, F. Kochan, L. Lunsford, N. Dominguez, & B. Smith (Eds.), *Sage handbook of mentoring* (pp. 316–334). Thousand Oaks: Sage Publications Ltd.

Marquis, E. (2017). Undergraduate research and student-staff partnerships: Supporting the development of student scholars at a Canadian teaching and learning institute. *Scholarship and Practice of Undergraduate Research Journal (SPUR), 1*(1), 39–44.

Mitacs. (2019). *About Mitacs.* Retrieved from: https://www.mitacs.ca/en/about

Naepi, S. (2019). Knowledge making: Indigenous undergraduate research as cultural and language revitalization. *Canadian Journal of Native Education, 41*(1), 85–102.

Naepi, S., & Airini. (2019). Knowledge Makers Increasing Indigenous Student Engagement. *Scholarship and Practice of Undergraduate Research, 2*(3), 52–60.

National Survey of Student Engagement. (2019). *NSSE 2017–18 high-impact practices: Canadian summary percentages by student characteristics.* Retrieved from: http://nsse.indiana.edu/2018_institutional_report/pdf/canada/NSSE17_18%20Canadian%20Grand%20HIP.pdf

Palmer, R., Hunt, A., Neal, M., & Wuetherick, B. (2015). Mentoring, undergraduate research and identity development: A conceptual framework and research agenda. *Mentoring and Tutoring, 23*(5), 411–426.

Palmer, R., Hunt, A., Neal, M., & Wuetherick, B. (2018a). Mentored undergraduate research: A multi-institutional investigation into students' perceptions of identity shifts in this learning sphere. In J. Moore, P. Miller, & M. Vandermaas-Peeler (Eds.), *Excellence in mentoring undergraduate research* (pp. 19–42). Washington, DC: Council for Undergraduate Research.

Palmer, R., Hunt, A., Neal, M., & Wuetherick, B. (2018b). A closer look at mentored undergraduate research: Students' perceptions of its influence on identity development. *Scholarship and Practice of Undergraduate Research Journal (SPUR), 2*(2), 4–14.

Poe, J., & Armstrong, D. (2019). Personal communications.

Robinson, F., Wuetherick, B., Martin, J., Strawson, C., Schmid, K., Greenwood, S., & Wolanski, N. (2006). Experiences in collaborative project-based study: There's a heifer in your tank. *NACTA Journal, 50*(4), 6–10.

Samarasekera, I. (2005). *Presidential inaugural address.* Edmonton: University of Alberta Office of the President.

Spronken-Smith, R., Brodeur, J., Kajaks, T., Luck, M., Myatt, P., Verburgh, A., Walkington, H., & Wuetherick, B. (2013). Completing the research cycle: A framework and strategies for promoting the student dissemination of research and inquiry. *Teaching and Learning Inquiry, 1*(2), 105–118.

Turner, N., Wuetherick, B., & Healey, M. (2008). International perspectives on student awareness, experiences and perceptions of research: Implications for academic developers in implementing research-based teaching and learning. *International Journal for Academic Development, 13*(3), 199–211.

Undergraduate Research Opportunities. (2019). *About URO at UBC.* Retrieved from: https://www.uroubc.com/

University of Alberta. (2019). *Research certificate in science.* Retrieved from: https://www.ualberta.ca/science/student-services/your-academics/certificates

University of Saskatchewan. (2018). *Annual report – Undergraduate research initiative.* Retrieved from: https://vpresearch.usask.ca/documents/students/undergraduate/Usask-UGR-Annual-Report-2018_final1.pdf

University of Saskatchewan. (2019a). *Collaboration in the community.* Retrieved from: https://vpresearch.usask.ca/students/undergraduate/spotlight-on-faculty/collaboration-in-the-community.php

University of Saskatchewan. (2019b). *First-Year Research Experience (FYRE).* Retrieved from: https://vpresearch.usask.ca/students/undergraduate/undergrad-research-opportunities/FYRE.php

Vajoczki, S. (2010). From the international desk: A Canadian perspective on undergraduate research. *Council on Undergraduate Research Quarterly, 31*(2), 41–47.

Wuetherick, B. (2007). *The integration of teaching and research in Canada: The undergraduate student perspective*. Proceedings of the international colloquium on academic enquiry, Marwell, 1–7.

Wuetherick, B., & McLaughlin, L. (2011). Students' perceptions of the learning environment: A partnership to enhance our understanding of the undergraduate experience. In S. Little (Ed.), *Staff-student partnerships in higher education* (pp. 185–200). London: Continuum International Publishing Group.

Wuetherick, B., Willison, J., & Shanahan, J. (2018). Mentored undergraduate research at scale: Undergraduate research in curriculum and as pedagogy. In J. Moore, P. Miller, & M. Vandermaas-Peeler (Eds.), *Excellence in mentoring undergraduate research* (pp. 181–202). Washington, DC: Council for Undergraduate Research.

CHAPTER 16

Critical Thinking and Scientific Integrity: Are University Students Ready Enough to Be Engaged?

Luisa Soares

Introduction

Successfully engaging in a high-quality curriculum is a crucial factor in students' academic success. The curriculum should include developing research skills, core scientific competencies, and ethical research practices. These are skills necessary for students to become contributing members of the scientific community. However, when arriving at the university, young people face many challenges. They need to adjust to new academic strategies and a different pace of work. They also must adapt to new teaching methodologies and evaluation; and more autonomy in studying and learning. On a personal level, it is necessary to strengthen the student's identity as an adult, confirm their commitment to a vocational pathway, and develop autonomy (Almeida et al. 2000). New patterns of interpersonal

L. Soares (✉)
Larsys – Interactive Technologies Institute, Madeira University, Funchai, Portugal
e-mail: lsoares17@hotmail.com

N. H. Hensel, P. Blessinger (eds.), *International Perspectives on Undergraduate Research*,
https://doi.org/10.1007/978-3-030-53559-9_16

relationships with family, peers, teachers, and other authority figures are additional areas requiring adaptation. It is not an easy developmental process.

University courses provide the framework for creative research activity. Still it is the interaction with faculty, peers, junior (undergraduate students), and senior researchers (masters and PhD students) that feed new students with active, inquiry-based learning experiences (Kinzie and Husic 2010). The university is a space that is considered a center of creation, transmission, and broadcasting of culture, science, and technology (Pacheco 2003). However, should we integrate undergraduate research into the university curriculum? Are undergraduate students mature enough to think, implement, and reflect on research results?

The Setting in Portugal

In Portugal, there are no organizations or government agencies that advocate for undergraduate research, unlike the availability for graduate research students from masters and PhD level degrees. The principal Portuguese public agency, Foundation for Science and Technology (FCT), had a more robust budget in 2019 when compared with previous years. For the year 2020, it will have €631 mn (an increase of 11%, €62 mn, compared to the year 2018). This budget includes national funds as well as funds from the European community. The foundation aspires to direct significant funds into scientific jobs. Their goal is to hire 5000 PhD researchers over the next three years. Other parts of the budget finance the national research centers, research and development of scientific culture (€138 million euros), and advance teaching (€114.5 mn) to improve education at all levels. To promote international cooperation in the area of science and technology, the FCT will manage €57 mn; scientific computation and access to scientific publications will manage €20.6 mn. Regarding the advanced teaching, the foundation funded more than 1600 PhD scholarships in 2019 (Firmino 2018). For an easy reference, the reader can check the FCT website that lays out FCT organization, priorities, and programs, as well as the definition of international cooperation (https://www.fct.pt/apoios/cooptrans/index.phtml.en).

Portugal has a population of around 10 million people, and students pay the university fee (about €900 per year). However, there is currently a broad discussion to provide free education for all students. While research-oriented curricula are well organized at doctoral and masters level, introduction to academic research at the undergraduate level is undeveloped in

Portugal, as it has not yet been considered from a national policy perspective. The main goal of the FCT budget is to foster research at the PhD level rather than the undergraduate level.

According to Kinzie and Husic (2010), undergraduate research, done well, engages multiple dimensions of a student's cognitive, behavioral, and attitudinal skills. Some examples include task-specific learning about instruments and methods that flow into active hypothesizing and procedural troubleshooting. The development of these skills may result in increased self-confidence and independence that help shape the student's vision for the future. The undergraduate research experience is enriched by attaining research experience early and often. This argument has been demonstrated empirically and discussed by Madan and Teltge (2013) and also throughout a variety of disciplines, including engineering (Narayanan 1999), medicine (Murdoch-Eaton et al. 2010), biology (Reynolds et al. 2009), physiology (Desai et al. 2008), neuroscience (Frantz et al. 2006), psychology (Wayment and Dickson 2008), as well as in multidisciplinary discussions in prestigious journals (e.g., Carrero-Martinez 2011; Russell et al. 2007).

Kinzie and Husic also argue that undergraduate student participation in research is now seen by many as a way of developing leaders for the twenty-first century. By presenting their research to campus-wide audiences, to peers at national conferences, to scientists at disciplinary society meetings, and legislators at the state and national levels, students learn to communicate at several levels—including "nonexpert" audiences (in terms of scientific literacy). In addition to improving students' communication skills, dissemination activities also enrich the public understanding of science and allow students to become ambassadors for illuminating the importance of science and research in society (Kinzie and Husic 2010; Electronic version).

Should Portugal Offer Undergraduate Research?

As a first thought, we might agree with Kinzie and Husic, but more reflection about the subject is needed. Undergraduate students should be educated to think critically and be cognitively mature, but it is only at the masters and PhD levels when real research projects should be implemented. The developmental age of university students ranging from 18 to 21/23 years should be considered, including cognitive and physical maturity. Although undergraduate students can fully understand abstract concepts and are aware of consequences and personal limitations, one has to consider the different rhythms of human development and that some students reach maturity earlier than others (Almeida et al. 2000, 2003; Albuquerque 2008).

The period of adolescence shows critical transformations on neural development through neurochemical and morphological aspects. In the last decade, neurobiological research has shown that the neurological maturity of adolescence ends at 25 years of age (expected age of masters' students). New technologies of images made by magnetic resonance, which weren´t available a few years ago, have allowed the new research to progress. The typical behaviors of adolescence are cognitive impulsivity, emotional instability, and the desire for dangerous situations (Andrade et al. 2018). Our knowledge about the neurobiology behind these cognitive and behavioral changes has increased significantly with the arrival of magnetic resonance imaging (MRI), which allows unprecedented access to the anatomy and physiology of the living brain. Longitudinal studies with MRI begin to map the developmental trajectories of brain maturation and to explore the genetic and environmental influences on these trajectories in health and disease. Among the findings, obtained by neuroimaging, the one that says that the prefrontal cortex (an essential component of the neural networks involved in judgment, decision making, and impulse control) continues its maturation when the person reaches 25 years of age, had a great influence in the social, legislative, judicial, parental, and educational fields (Giedd 2011). Executive cognitive functions are processes that support many daily activities, including planning, flexible reasoning, focused attention, and behavioral inhibition, and demonstrate continuous development until early adulthood (Knapp and Morten 2013). A critical perspective for the development of these psychological skills is the structural and functional development of the brain, and one of the slowest developing brain regions is the prefrontal cortex, a large extension of the cortex located in the frontal half of the brain. What is remarkable about this region of the brain is that it continues to develop until the third decade of life (Knapp and Morten 2013).

Entry into higher education is a significant transition in a student's life, as it represents the possibility of continuing personal and professional projects. The student usually arrives at this educational level with strong expectations about the nature and characteristics of the academic context (Coelho et al. 2014). According to the literature (Albuquerque 2008; Almeida et al. 2003), it is during the first year of undergraduate courses when expectations about the university seem to dawn upon students, and significant adaptation difficulties appear, possibly resulting in academic failure. According to Almeida et al. (2000), the transition requires overcoming academic, personal, social, and vocational challenges. A study conducted by Fernandes et al. (2005) with 48 newly arrived students to the University of Minho, in the north of Portugal, revealed that many students evidence dilemmas on entry to higher education as well as severe symptoms that may reflect maladjustment and psychological distress associated with the challenging period and its new demands. So, teaching students how to do research should be designed at different developmental levels, regarding first-year students and senior students, like masters or PhD students. This is the scenario in Portugal, and I have no reason to believe teenage students' cognitive maturity should be so much different in other countries.

Universities should intentionally promote critical thinking in all science courses, aiming to achieve scientific integrity for students in the future. Building an excellent educational and scientific core foundation, providing space for growth, helping students to mature into adulthood with a stable psychological balance, and respecting the normal immaturity of university students, is more likely to develop competent researchers. How can we promote this growing process in students?

The Plea for Scientific Literacy: Connections Between Scientific Literacy, Scientific Thinking, and Critical Thinking

Faculty and mentors ought to encourage scientific and critical thinking in students, but how to do this? To think critically requires clarity, solid arguments, rigorous and systematic reasoning, based on scientific evidence. Melo (2016) suggests the following considerations might help teachers to develop critical thinking skills in students since teachers have a significant role in activating and modifying this critical thinking:

1. Identify fallacies in students' arguments, like wrong reasoning that nonetheless seems plausible.
2. Watch out for superstitions from students. Avoid assuming that events that happen sequentially have a causality relationship, before

forming conclusions, run out all possibilities from a situation. Know that coincidences do happen.

3. Study reliable and safe sources like scientific journals with high credibility.
4. Present graphics and statistics that are incomplete, regarding their sources and the percentage of the sample. Ask students to identify the problems.
5. Be intellectually honest and exude scientific integrity. Be accurate on acquiring, transmitting, and analyzing ideas. Don't make the mistake of hiding or demurring information just so that they corroborate your beliefs. Value the divergent results from a study. Arguments in favor and against should be analyzed rigorously and with no partiality involved. Honesty is essential so that science and the debate around it move forward. Go back to basics on this. If you think students already know that, it is never too much to mention this; especially in these fragile times that we are living in, concerning respect, values, morality, and honesty.
6. Have an open mind to new ideas, be ready to discuss the old ones, and look for different perspectives on the same subject. On evidence, be brave to change your opinion and admit it; that is being a good role model for students.
7. Promote questioning at every moment: introduce exciting questions throughout the classes and relate them to students' everyday life; you need to know their reality and know how to deal with problems when they occur.
8. Raise polemic themes that require students' positioning with arguments; facilitate them with respect and tolerance for different opinions.
9. Even in expositive classes, questioning and dialogue should be stimulated.
10. Help students to develop their ideas and construct their arguments in a logical and organized manner, giving them positive-negative-positive feedback. Teach students to build self-critical capacity. This is a significant skill of critical thinking, so it is essential to motivate students to analyze their speech, their arguments, guiding them when inconsistencies and incoherent sentences are found. This is when you can help to reassess some positioning and eventually change opinions/arguments, encourage the formation of new

positioning based on scientific evidence and not on preconceptions and biases.

11. Assist students in maintaining the focus on a subject, avoiding escapes from the theme being discussed.
12. Without venerating the Internet as the solution for the educational problems, contents from YouTube and TedTalks, for instance, if cleverly selected, can provide different points of view about a subject, increasing the debate and discussion in classrooms. For example, using YouTube videos about ethical and real problems and TedTalks can present a different point of view on a subject.

Let us now look at the definition of scientific integrity presented by Inserm (La science pour la santé/from science to health) by a team of researchers in France (Inserm 2019):

> *Scientific integrity is the truthful and honest conduct that must govern all research. Inherent to all research activities, it forms the basis of knowledge and learning. Scientific integrity is not a moral issue but is founded on universal moral principles according to which it is wrong "to lie, to steal,* etc.*" The quality and reliability of scientific output depend on it. The knowledge society is founded on it to—to put it succinctly—"believe in science." While ethical issues are debated, scientific integrity is indisputable. Scientific integrity is self-respecting; it is a code of professional conduct that must not be infringed. It is essential for science, in the same way as the professional codes of ethics are crucial for medicine and law.*

Research in the university curriculum should be organized for different levels of complexity in learning. As I see it in Portugal, the first three years of formation (undergraduate students) should include education in critical thinking and scientific integrity, parallel to the specific contents of each course. Building an ethical balance between the notions of personal gain versus common gain, educating for balance between rights and duties is a goal. If students have a solid education on this combination between rights and responsibilities toward research and society, they won't have ethical issues regarding the development of quality research results. If these notions are firmly developed in their curriculum, we can emphasize opportunities for observing, with a mentor, the research into real contexts where real problems occur that need solutions.

I agree with Kinzie and Husic (2010) when they suggest engaging students in the collection and analysis of original data with mentor supervision increases students' ownership of the research project over time, but only at the masters' level, not undergraduate. At the masters' level, students should be cognitively and emotionally mature enough to engage in real substantive matters with a solid bioethical foundation.

Professor Sally Hoskins (National Science Foundation support in USA 2003) presents a very interesting approach to develop undergraduate research with the method C.R.E.A.T.E. The C.R.E.A.T.E. (Consider, Read, Elucidate the hypotheses, Analyze and interpret the data, and Think of the next Experiment) method is a new teaching approach that uses intensive analysis of primary literature to demystify and humanize research science for undergraduates. The teaching/learning strategy developed and expanded in the United States with National Science Foundation support (2003–present) promotes the development of transferable analytical skills by focusing in-depth on a series of papers from a single research group. C.R.E.A.T.E. builds students' critical thinking and content integration abilities at the same time that it transforms their understanding of the research process and aspects of their epistemological beliefs (National Science Foundation support in USA 2003).

According to Kinzie and Huzic, a white paper published by The Teagle Foundation Working Group on the Teacher–Scholar (2007) approach, provides a concise argument for the robust connections and synergy between teaching and scholarship, at both undergraduate institutions and research universities. Fully embracing the pedagogy of discovery, inquiry, and analysis suggests the integration of teaching with research as opposed to separation. I agree with this argument. Still, it should not be applied as a real-world problem-solving project at the undergraduate level, only at the level of master's degree, as the Bologna Declaration upholds since 1999.

Imagine this scenario with three levels of research learning: students from the 1st cycle are juniors in research and observe the master students, learning by observation on how to design and implement research. And both these students work in cooperation with PhD students who are in real contact with real problems. The goal is to develop research skills and expand student mindsets to science and research, with a solid foundation on bioethical principles and supervised by senior students and faculty members. It not only enhances the problem-solving and analytical skills of the students but also promotes collaboration and teamwork among them

(Hati and Bhattacharyya 2018). The symbiosis established between students introduces them to the joys of discovery as well as lessons in persistence, problem-solving, and critical thinking (Kinzie and Husic 2010).

Education in the twenty-first century must be updated, and we need to look for an understanding that transcends the classroom, but there is still a lot to be done inside the classroom. In particular, we need to find a balance between practical and theoretical concepts.

Worldwide, there are complex and global challenges like climate change, energy usability, and especially world health in elderly people. Also, on another spectrum of reality, as Damásio (2019) stated, social networks are disturbing the political process in a considerable way, such as the rapid and massive access to bad information (not well analyzed) is a considerable risk. Human life will be less respected when considering human values. These challenges will most certainly involve multidisciplinary perspectives that need many soft skills like finding common ground of communicating with different scientific areas, managing different arguments, mediating different personalities and, most importantly, supporting and representing scientific integrity within a culture of reliability, conducting responsible research. Kretser et al. (2019) present the result of a scientific integrity consortium, where they developed a set of recommended principles and best practices that could be used broadly across scientific disciplines as a mechanism for consensus on scientific integrity standards. The authors (Kretser et al. 2019) present two main principles under which scientific processes should operate: (1) foster a culture of integrity in the scientific method and (2) evidence-based policy interests may have legitimate roles to play in influencing aspects of the research process. Still, those roles should not interfere with scientific integrity.

Critical Thinking and Scientific Integrity: Are Undergraduate Students Ready to Adopt It?

Yes, but let's help them grow first. Kretser et al. (2019) argue about nine best practices for instilling scientific integrity:

I. Require universal training in robust scientific methods, in the use of appropriate experimental design and statistics, and in responsible research practices for scientists at all levels, with the training content regularly updated and presented by qualified researchers.

II. Strengthen scientific integrity oversight and processes throughout the research continuum with a focus on training in ethics and conduct.
III. Encourage reproducibility of research through transparency.
IV. Strive to establish open science as the standard operating procedure through the scientific enterprise.
V. Develop and implement educational tools to teach communication skills that uphold scientific integrity.
VI. Strive to identify ways to strengthen the peer review process further.
VII. Encourage scientific journals to publish unanticipated findings that meet standards of quality and scientific integrity.
VIII. Seek harmonization and implementation among journals of rapid consistent and transparent processes for correction or retraction of published papers.
IX. Design rigorous and comprehensive evaluation criteria that recognize and reward the highest standards of integrity in scientific research.

Should we demand that new undergraduate students, facing so many personal developmental challenges, be ready to think, implement, and analyze in a mature and scientific way, research results? Are they prepared to face our global society, communicate science to decision-makers? Although some communities of researchers have proposed that early undergraduate research experiences that are grounded in incoming students' experiences and communities are effective means of student growth and development, it seems wiser to implement first, a culture of scientific integrity, and then demand quality and a scientific attitude from students, concerning the design, implementation, and examination of scientific results. Studies of Vygotsky (1981), about development and learning processes, have made an essential contribution to the educational field. All the initial learning processes benefit from significant social experiences, with the mediation of other persons with whom the individual can interact, whether adults or peers. This is very important so that the learning of new concepts can happen smoothly and regularly. Let's not rush the development of university students. The world has considerable problems to be addressed. Political and social issues are directly related to science and research and, in that sense, an overview of considerations about struggles in Portugal, Europe, and the world is presented next in order to conclude why it is so important to have a solid education on ethics and research principles.

Portugal is part of the European community, and the European integration has operated, during the past decades, as a safe harbor of peace, respect for human rights, defense of democracy, endowing well-being, supporting the free circulation of citizens, analogous to opening frontiers to hospitality and receiving citizens from third-world countries. All this occurred on behalf of the agreement of Schengen (Official Journal of the European Communities 2000). The European community has also been, since its foundation, a space of construction of shared citizenship and development of global ecological consciousness (Carta Pastoral 2019). We are facing global problems like biological diversity, climate variability, health issues, elderly health care, social migration, and social integration. At this moment, we are witnessing movements of disintegration in the European Union, like Brexit (i.e., the exit of the United Kingdom out of the European Union, cf. Wikipedia for more political detailed information) and the increase of national authoritarian movements, parallel to a rise of xenophobic statements. The intolerance toward different cultures seems to be increasing. Scientific teams from all fields should embrace this problem genuinely when supporting the fact that students from master or PhD courses present an attitude of scientific integrity. Undergraduate students should be part of these research teams, observing the senior's approach. These teams should promote solutions that contest an environment of physical, verbal, and psychological violence, and promote, instead, cooperation, solidarity, and economic and social guidelines. In particular, science should mainly aim to increase social cohesion between different social classes and support their findings keeping in mind the sustainability of a more fair-minded society. A study from Organization to Cooperation and Economic Development (OCDE), entitled *In it* together—why less *inequality benefits all* (May 2015) shows how the levels of inequality are the highest in the last 30 years. In the 1980s, in the last century, the proportion of resources from the 10% wealthiest people and the 10% poorest, was from 1 to 7. Today is from 1 to 10. These issues demand cognitive and emotional maturity and the scientific integrity of students/researchers. Science goals, in all scientific fields, should aim to help human beings and target for a better future, where research communities contribute to a more nondiscriminatory and just society.

References

Albuquerque, T. (2008). Do abandono à permanência no ensino superior. *Revista Ciências da Educação, 7*(1), 19–28. Available at http://dialnet.unirioja.es/servlet/articulocodigo1/42795158.4

Almeida, L., Soares, A., & Ferreira, A. (2000). Transição e adaptação à Universidade: Apresentação de um questionário de vivências académicas (QVA). *Psicologia, 14*(2), 189–208. Available at http://repositorium.sdum.uminho.pt/handle/1822/12069

Almeida, L. S., Gonçalves, A., Salgueira, A. P., Soares, A. P., Machado, C., Fernandes, E., Machado, J. C., & Vasconcelos, R. (2003). Expectativas de envolvimento académico à entrada da Universidade: Estudo com alunos da Universidade do Minho, *Psicologia: Teoria. Investigação e Prática, 8*(1), 3–15.

American Council of Learned Societies. (2007, May). *Student learning and faculty research: Connecting teaching and scholarship*, A Teagle Foundation white paper.

Andrade, A. L. M., Bedendo, A., Enumo, S. R. F., & Micheli, D. (2018). Desenvolvimento cerebral na adolescência: aspectos gerais e atualizaçao. *Adolesc Saude, 15*(Supl. 1), 62–67. https://s3-saeast1.amazonaws.com/publisher.gn1.com.br/adolescenciaesaude.com/pdf/v15s1a08.pdf

Carrero-Martinez, F. A. (2011). Rethink summer student research. *Science, 334*, 313.

Carta Pastoral. (2019). Um olhar sobre Portugal e a Europa à luz da doutrina social da Igreja. In *Conferência Episcopal Portuguesa*, Fátima, Portugal.

Coelho, C., Faria, C., Oliveira, F., Lucas, C. V., Vasconcelos, S., & Soares, L. (2014). Pilot project for peer tutoring: A case study in a Portuguese university. *Journal of Applied Research in Higher Education, 6*(2), 314–324. ISSN 2050-7003.

Damásio, A. (2019, June 8). Interview to journal *Expresso.* Retrieved on December 13, from https://expresso.pt/sociedade/2019-06-08-As-redes-sociais-estao-a-prejudicar-a-qualidade-de-vida.-Palavra-do-neurocientista-Antonio-Damasio?fbclid=IwAR0oO2Q83Gl0Y4z_tDbb9Se23F02Cu-y-kPWBF8qSpXbd_tWsSVlaP1Tbks

Desai, K. V., Gatson, S. N., Stiles, T. W., Stewart, R. H., Laine, G. A., & Quick, C. M. (2008). Integrating research and education at research-extensive universities with research-intensive communities. *Advances in Physiology Education, 32*, 136–141.

Fernandes, E., Maia, A., Meireles, C., Rios, S., Silva, D., & Feixas, G. (2005). Dilemas implicativose ajustamento psicológico: um estudo com alunos recém-chegados à Universida deMinho. *International Journal of Clinical and Health Psychology, 5*(2), 285–304.

Firmino, T. (2018). FCT vai ter um dos maiores orçamentos de sempre. *Jornal o Público.* Retrieve https://www.publico.pt/2018/10/16/ciencia/noticia/fct-vai-ter-um-dos-seus-maiores-orcamentos-de-sempre-1847676.

Frantz, K. J., DeHaan, R. L., Demetrikopoulos, M. K., & Carruth, L. L. (2006). Routes to research for novice undergraduate neuroscientists. *CBE Life Sciences Education, 5*, 175–187.

Giedd, J. N. (2011). Maturação do cérebro adolescente. In *Enciclopedia sobre desenvolvimento da primeira Infância*, Eletronic version. Retrieved on March 16, 2020, from http://www.enciclopedia-crianca.com/cerebro/segundo-especialistas/maturacao-do-cerebro-adolescente

Hati, S., & Bhattacharyya, S. (2018). Integrating research into the curriculum: A low-cost strategy for promoting undergraduate research. In *Best practices for supporting and expanding undergraduate research in chemistry, ACS symposium series, 8*(1275), 119–141, ISBN13: 9780841232846, ISBN: 9780841232839. https://doi.org/10.1021/bk-2018-1275.ch008.

Inserm. (2019). La science pour la santé/From science to health. *Assessment and proposals for the implementation of the national charter of scientific integrity*, by Pierre Corvol with the participation of Rémy Giscquel (2016 in French). Retrieved on July 29, 2019, from https://www.inserm.fr/en/research-inserm/scientific-integrity

Kinzie, J., & Husic, D. (2010). Research and discovery across the curriculum. *Peer Review*, 12, 2. Retrieved July 16, 2019, from (Electronic version) https://www.aacu.org/publications-research/periodicals/research-and-discovery-across-curriculum

Knapp, K., & Morten, B. J. (2013). Desenvolvimento do Cérebro e Funcionamento Executivo. In *Enciclopedia sobre desenvolvimento da primeira Infância*, Eletronic version. Retrieved on March 19, 2020, from http://www.enciclopedia-crianca.com/funcoes-executivas/segundo-especialistas/desenvolvimento-do-cerebro-e-funcionamento-executivo

Kretser, A., Murphy, D., Bertuzzi, S., Abraham, T., Allison, D. B., Boor, K. J., Dwyer, J., Grantham, A., Harris, L. J., Hollander, R., Jacobs-Young, C., Rovito, S., Vafiadis, D., Woteki, C., Wyndham, J., & Yada, R. (2019). Scientific integrity principles and best practices: Recommendations from a scientific integrity consortium. *Science and Engineering Ethics, 355*(25), 327. https://doi.org/10.1007/s11948-019-00094-3.

Madan, C. R., & Teltge, B. (2013). *The benefits of undergraduate research: The student's perspective, 15*, electronic version. https://doi.org/10.26209/MJ1561274. Retrieved December 18, from https://journals.psu.edu/mentor/article/view/61274/60907

Melo, A. S. (2016). Seis maneiras de estimular o pensamento critico. *Universo Racionalista*. Retrieved December 13, from https://universoracionalista.org/seis-maneiras-de-estimular-o-pensamento-critico/

Murdoch-Eaton, D., Drewery, S., Elton, S., Emmerson, C., Marshall, M., Smith, J. A., & Whittle, S. (2010). What do medical students understand by research and research skills? Identifying research opportunities within undergraduate projects. *Medical Teacher, 32*, e152–e160.

Narayanan, R. M. (1999). Use of objective-based undergraduate research project experience as a graduate student recruitment tool. *Journal of Engineering Education, 88*, 361–365.

National Science Foundation support USA. (2003–Present). *C.R.E.A.T.E.* (*Consider, Read, Elucidate the hypotheses. Analyze and interpret the data, and Think of the next Experiment*, July 2019). Retrieved on July 18, 2019, from http://www2.cuny.edu/research/student-resources/for-faculty/integrating-research-into-the-curriculum/

OECD. (2015). *In it together: Why less inequality benefits all.* Paris: OECD Publishing. https://doi.org/10.1787/9789264235120-en.

Pacheco, J. A. (2003). Políticas educativas para o ensino superior na União Europeia: Um olhar do lado português. *Educação e Sociedade, 24*(82), 17–36.

Reynolds, J., Smith, R., Moskovitz, C., & Sayle, A. (2009). BIOTAP: A systematic approach to teaching scientific writing and evaluating undergraduate theses. *Bioscience, 59*, 869–903.

Russell, S. H., Hancock, M. P., & McCullough, J. (2007). Benefits of undergraduate research experiences. *Science, 316*, 548–549.

Schengen Agreement. (2000). Convention implementing the schengen agreement of 14 June 1985. *Official Journal of the European Communities*, 19–62. Retrieved on December 13, 2019, from https://eur-lex.europa.eu/legal-content/EN/TXT/PDF/?uri=CELEX:42000A0922(02)&from=EN

Vygotsky, L. (1981). The genesis of higher mental functions. In J. V. Wersch (Ed.), *The concept of activity in soviet psychology* (pp. 144–148). Armonk: M.E. Sharpe.

Wayment, H. A., & Dickson, K. L. (2008). Increasing student participation in undergraduate research benefits students, faculty, and department. *Teaching of Psychology, 35*, 194–197.

CHAPTER 17

Conclusions and Recommendations

Nancy H. Hensel and Patrick Blessinger

Engaging students in research is an educational approach that is at least 200 years old. As Mieg and Ambos mentioned in their chapters, the roots of undergraduate research can be traced back to the founding of the Humboldt University of Berlin by Wilhelm von Humboldt in 1810. Humboldt suggested that the *pursuit of knowledge through original research and the partnership of student and professor* is the core mission of the university (Zupanc 2012). Undergraduate research became a more significant part of undergraduate students' education in the United States after the founding of the Council on Undergraduate Research (CUR) in 1978. As CUR grew in membership and influence, faculty from several other countries reached out to CUR, and some attended CUR conferences and workshops. IN 2010, CUR created the international desk for the CUR Quarterly. Mick Healey and Alan Jenkins, from Great Britain,

N. H. Hensel (✉)
Laguna Woods, CA, USA
e-mail: hensel.nancy@gmail.com

P. Blessinger
International Higher Education Teaching and Learning Association,
New York, NY, USA
e-mail: patrickblessinger@gmail.com

N. H. Hensel, P. Blessinger (eds.), *International Perspectives on Undergraduate Research*,
https://doi.org/10.1007/978-3-030-53559-9_17

were invited to be the first international editors. Healey and Jenkins had many contacts with faculty around the world who were implementing inquiry-based teaching. Articles from the CUR International Desk suggested a need for a book about international undergraduate research.

The authors included in this book have described the status, challenges, and successes of undergraduate research in their countries. There are many differences in how undergraduate research is defined, the resources available for student research, and the implementation at colleges and universities.

National Organization or Government Agency Support

The United States is the only country among those included in this book that has a national organization with full-time paid staff and a physical office. The CUR office, in Washington D.C., has an executive director and several staff members to handle membership, communication, event planning, finances, and the student conference, National Conferences on Undergraduate Research (NCUR). CUR also has a government relations firm that advocates with the US Congress for the inclusion of undergraduate research in federal funding through various agencies such as the National Science Foundation, Department of Education, Department of Energy, National Endowment for the Humanities, and other federal agencies. The government relations firm also keeps CUR members informed about funding issues through a monthly newsletter. In 2010, the US House of representatives authorized an Undergraduate Research Week that is now celebrated annually by many campuses across the country. CUR hosts an annual "Posters on the Hill" at the US Capitol, its major advocacy event, and Members of Congress attend. The National Science Foundation supported several CUR projects to support faculty development to engage students in research. CUR also has published nearly 20 books and sponsors a quarterly journal. The national CUR office and all of the activities it offers have given undergraduate research national and international visibility. Undergraduate research in the United States would not be where it is today, were it not for the vision of the CUR founders and those who expanded the original idea and mission. CUR now has over 10,000 members and about 1000 institutional members and can effectively advocate for research funding and policies because of the strength of its membership. Membership includes community colleges, liberal arts colleges, and research-intensive universities and all disciplines.

The British Conference of Undergraduate Research (BCUR) was founded in 2010 to promote undergraduate research in all disciplines. It hosts an annual conference for students and accepts proposals from students outside of the United Kingdom. BCUR also organizes a "Posters at Parliament" each year as part of the annual meeting. The conference has been growing every year since its inception in 2011. "Posters at Parliament," inspired by CUR's "Posters on the Hill," is the primary advocacy event for BCUR. It allows Members of Parliament and other policymakers an opportunity to see the excellent work of students and develop a deeper understanding of the research efforts of British universities.

Australia also has a national organization, Australasian Conference of Undergraduate Research (ACUR), founded as a nonprofit organization in 2012. It supports three main events. The primary event is the Australasian Conference of Undergraduate Research for students from Australian and New Zealand universities. ACUR also hosts an annual "Posters in Parliament" and workshops and summits for faculty.

Another interesting conference is the International Conference on Undergraduate Research. It is a collaborative effort between Monash University in Australia and Warwick University in Great Britain, known as the Monash Warwick Alliance. The conference encourages students to think about their work from an international and interdisciplinary perspective. Students study global, regional, and local trends in their research field and look for connections between disciplines. Research projects are presented in real-time, video-linked sessions. Only students at institutions that belong to the Monash Warwick Alliance can participate in ICUR. The World Congress on Undergraduate Research is the newest international conference. CUR, BCUR, ACUR, and Qatar University hosted the first World Congress. A result of the Congress was the establishment of the Alliance for Global Undergraduate Research. Partners in the Alliance are ACUR, BCUR, CUR, and Qatar University. Others may eventually join the Alliance to engage in planning and hosting the conferences.

It is interesting to note that the main activity of ACUR, BCUR, and ICUR is a conference for students. The meetings are an important vehicle for promoting undergraduate research in their respective countries and faculty in those countries actively participate in organizing and hosting the gatherings. These three organizations may eventually evolve into a full-fledged organization similar to CUR or they may remain as supporters of a major conference and other professional development activities. Australia and Great Britain have more robust undergraduate research programs than many of the other countries represented in this book. The annual conferences have given visibility to their work and are undoubtedly a factor in the expansion of undergraduate research.

Definitions of Undergraduate Research

There are many different definitions of undergraduate research, and in most of the countries included in this project, there is no universally adopted definition. The CUR definition perhaps comes closest to being adopted by a large number of higher education institutions:

> *An inquiry or investigation conducted by an undergraduate student that makes an original intellectual or creative contribution to the discipline.*

Widely used, the CUR definition has several problems. First, many feel that the emphasis on an original contribution is a very high bar for undergraduate students to achieve. In the United States, community colleges often take issue with the originality requirement. Faculty who teach students in their first two years often say that students do not know enough or have enough experience to make an original contribution. When I spoke about undergraduate research as executive director of CUR, I usually said that the CUR definition is the gold standard and that it is what we ultimately aspire to for our students. We do not expect beginning students to do groundbreaking research; however, we can prepare them to do more advanced research by scaffolding the necessary skills of research in the discipline. The United States places significant emphasis on the product of research and the communication of the results even for beginning research efforts. Bellevue Community College, for example, has a robust undergraduate research program. They introduce biology students to research that leads to a discovery. The research skills are carefully scaffolded with students first doing a literature review and ending with DNA analysis and identifying new bacteria. While the students make an original contribution on a minor scale, the biology department felt a need to revise the CUR definition of research. Their definition reflects the level of student development and appropriate expectations for first-year students:

> *Research is a purposefully structured activity that results in the generation of noteworthy data that did not exist before, the analysis of that data leading to specific inferences, conclusions, and making these results available to a community of scientists able to take advantage of those data.* (Hensel 2019)

The Bellevue definition, while developed for science students, is broad enough that it could be applied to beginning students in other disciplines as well.

I was part of a CUR team that offered a workshop in Canada as part of the Carnegie Academy for the Scholarship of Teaching and Learning. As team members, we were surprised by the Canadian discussion about the definition of research. Workshop participants expressed concern about the emphasis on the product of research. They suggested that they were more comfortable with a focus on the process of research and inquiry-based learning (Beckman and Hensel 2009). Many of the authors included in this book described similar views. The UAE is the exception with its focus on the development of the nation and the expectation that research would lead to solutions to societal and economic issues. Mandla S. Makhanya suggests that undergraduate and graduate research plays a role in meeting international goals for sustainability.

Several authors in this book used the term URI, undergraduate research and inquiry, which includes a range of learning experiences that develop the skills and attitudes described by many scholars who have studied the benefits of undergraduate research. Brew and Mantai expanded the CUR definition of undergraduate research to include *a research-based activity* that contributes *to understanding* as well as an original contribution (see p. 4).

Basic Models of Undergraduate Research

Ambos identified four primary models of undergraduate research: the apprentice model; research embedded in the curriculum; community-based research; and partnerships with businesses, research laboratories, and agencies. The apprentice model is very common and can be found on most campuses that support undergraduate research. Undergraduate research began with the apprentice model in the sciences where faculty would work with three or four students in their laboratory. This model can be found in every country that is represented in this book. The apprentice model provides a rich experience for the few students who are able to work closely with a faculty member. The apprentice model, however, is expensive and by necessity, the number of students who can take advantage of the program is limited. By extension, the apprentice model limits the expansion of undergraduate research.

Research embedded in the curriculum may be a newer trend in higher education and is a way to provide more students the opportunity to engage in undergraduate research. Biology programs in the United States are actively adopting course-based undergraduate research known as CUREs (Curriculum-based Undergraduate Research Experiences). There are many approaches to CUREs but generally it involves all students in a class who participate in a collaborative research project. In Germany, Deicke and Mieg found that the social sciences and humanities are more likely to embed research in their courses. The ability to reach more students is a clear advantage of course-based research. Making research available to more students is a strategy to make the college experience more equitable for all students.

Community-based research typically involves students working with a community agency such as the municipal water district or local historical society on a project that students may do either individually or collaboratively. Elshimi found that one Egyptian university focused on the community-based model and saw research as a civic skill developed to build and support the local community and environment. Several authors mentioned the importance of partnerships with other universities, research laboratories, and industry. International partnerships are becoming a way to expand and enhance research opportunities for students.

Students as Producers, Faculty as Mentors

Brew suggests that undergraduate research is transformative because assumptions about students, the nature of knowledge, and who produces it must be examined. Students and faculty collaborate to produce new knowledge and the transition can be challenging for both. Some faculty have a difficult time giving up their role as the ultimate authority in the classroom or laboratory. And some students are intimidated by the idea that they need to go beyond learning facts and think critically about issues and solve problems. Wuetherick discussed a Canadian initiative called *Knowledge Makers* where students work collaboratively with researchers and Indigenous elders to integrate Western knowledge and Indigenous knowledge.

Honoring indigenous knowledge creates a welcoming environment for minoritized students. Finding ways to involve students in the design and implementation of research recognizes the value of multiple cultural experiences and leads to a more equitable environment.

Funding Undergraduate Research

Insufficient funds are a challenge that nearly every institution and country must address. Building closer connections between university undergraduate research programs and community agencies, local businesses and industries can provide a more stable base for fiscal support. Presenting student research to regional or national legislative bodies helps those who control funding to see the excellent and sophisticated work of undergraduate students. Some countries do have governmental support for student research through agencies like the National Foundation for Science in the United States or the National Council for Scientific and Technological Development, Brazil.

Undergraduate Research as Workforce Preparation

Several authors discussed the tension between the emphasis on scholarship and career preparation. Many in the academy resist the idea of a direct link between an undergraduate's course of study and preparing for a future career. Studies in the United States, however, have found that the skills developed through undergraduate research are often the skills that prepare students for success in future careers. The Association of American Colleges & Universities collaborates each year with Hart Associates to conduct a survey of what employers look for in future employees (2018). The 2018 survey found, for example, that employers want students with the following skills:

- Oral and written communication skills
- Critical/analytical reasoning skills
- Ethical judgment/decision making
- Ability to work in teams and independently
- Self-motivated/initiative proactive
- Ability to apply skills to real-world problems

These are skills that we hope our graduates will develop regardless of major or future career. They are also skills that can be developed through undergraduate research experiences.

Recommendations

Drawing from the chapters included in this book, the authors suggest the following recommendations to further support undergraduate research:

- A national organization can help to advance undergraduate research by bringing attention to student research through conferences, workshops, and advocacy events such as BCUR's "Posters in Parliament" or CUR's "Posters in the Capitol."
- Definitions of undergraduate research need to be particular to the culture of the country, the university, and the discipline.
- Consistent administrative support is needed to develop and sustain undergraduate research.
- Faculty need professional development opportunities such as the modules developed by Roisin Donnelly and colleagues.

The skills that students learn through participation in undergraduate research, such as critical and analytical thinking, ethical judgment, and use of evidence, are skills that will benefit students regardless of the career they enter. They will be better citizens when they can ask questions and analyze information. Students who can think critically can more effectively contribute to addressing local issues and developing solutions for national and international problems.

References

Beckman, M., & Hensel, N. (2009). Making explicit the implicit: Defining undergraduate research. *CUR Quarterly, 29*(4), 40–44.

Hart Research Associates. (2018). *Fulfilling the American dream: Liberal education and the future of work*. Washington, DC: Association of American Colleges and Universities. https://www.aacu.org/research/2018-future-of-work

Hensel, N. (2019). *Course-based undergraduate research: Educational equity and high impact practice*. Sterling: Stylus Publishing, LLC.

Zupanc, G. K. H. (2012). Undergraduate research and inquiry-based learning: The revitalization of the Humboldtian ideals. *Bioscience education, 19*(1), 6.

Index

N. H. Hensel, P. Blessinger (eds.), *International Perspectives on Undergraduate Research*,
https://doi.org/10.1007/978-3-030-53559-9

The manufacturer's authorised representative in the EU is Springer Nature Customer Service Centre GmbH, Europaplatz 3, 69115 Heidelberg, Germany. If you have any concerns regarding our products, please contact ProductSafety@springernature.com

Printed and bound by CPI Group (UK) Ltd, Croydon, CR0 4YY
30/09/2026
02225456-0009